Essentials of Bioinformatics, Volume III

Khalid Rehman Hakeem • Noor Ahmad Shaik
Babajan Banaganapalli • Ramu Elango
Editors

Essentials of Bioinformatics, Volume III

In Silico Life Sciences: Agriculture

 Springer

Editors
Khalid Rehman Hakeem
Department of Biological Sciences
King Abdulaziz University
Jeddah, Saudi Arabia

Princess Dr. Najla Bint Saud Al-Saud
Center for Excellence Research
in Biotechnology
King Abdulaziz University
Jeddah, Saudi Arabia

Babajan Banaganapalli
Princess Al-Jawhara Center of Excellence
in Research of Hereditary Disorders
Department of Genetic Medicine
Faculty of Medicine
King Abdulaziz University
Jeddah, Saudi Arabia

Noor Ahmad Shaik
Princess Al-Jawhara Center of Excellence
in Research of Hereditary Disorders
Department of Genetic Medicine
Faculty of Medicine
King Abdulaziz University
Jeddah, Saudi Arabia

Ramu Elango
Princess Al-Jawhara Center of Excellence
in Research of Hereditary Disorders
Department of Genetic Medicine
Faculty of Medicine
King Abdulaziz University
Jeddah, Saudi Arabia

ISBN 978-3-030-19317-1   ISBN 978-3-030-19318-8   (eBook)
https://doi.org/10.1007/978-3-030-19318-8

This Springer imprint is published by the registered company Springer Nature Switzerland AG.
The registered company address is: Gewerbestrasse 11, 6330 Cham, Switzerland

Foreword

Bioinformatics is an interdisciplinary area of study and merges the various fields of biological sciences, mathematics, computers, statistics, physics, and chemistry. It is essentially used to develop methods, tools, and databases to analyze and interpret the biological data, which has seen a steep increase in the last two decades. The upsurge of this biological data is attributed to the expeditious advancements in sequencing, microarray, and proteomic technologies. This data has huge potential to serve as a basic platform for biological knowledge discovery, possible only with the help of advanced computational methods that are capable to tackle the data of this magnitude.

Plants have huge impact on our lives as well as the global environment. With the accumulation of diverse types of data from plants with agricultural implications, some of the main challenges in agricultural sciences are to improve the crop productivity and predict the response of plants under stressful environmental conditions. The voluminous quantity of data generated from multi-omics studies offers potential in achieving these objectives by understanding the functionality of the agriculturally important crops. The increase in the amount of biological data from plants has also seen a spectacular emergence in the development of plant-specific databases and tools. These bioinformatics resources enable the researchers to investigate the molecular differences in plants at various stages and conditions at a large scale. This would in turn lead to necessary genome modifications in the plants of agricultural significance with a possible improved productivity. Other key elements in regulating the overall development of plants are the microbes associated with them. With the advancements in metagenomics, it is now possible to profile the taxonomic composition and functional roles of these microbial communities, thereby providing opportunities for the discovery of novel genes, secondary metabolites, or even novel microorganisms that may affect plant growth or productivity. Each step aimed at improving the plant productivity will witness a crucial contribution from bioinformatics. Therefore, bioinformatics will not only be an important tool in agricultural research, but it will serve as a major driving force for designing future studies in this leading field of plant science.

This book highlights some of the major areas of agricultural research in which bioinformatics has been applied to address certain specific questions. For example, in addition to the topics related to crop productivity, stress, and heavy metal tolerance, the book also focusses on the recent trends in medicinal plant research, evolutionary analysis, and sequencing technologies. The book is of interest to the students as well as researchers with deep interest in both agricultural sciences and bioinformatics who either employ or develop these computational methods for better understanding of the resultant data.

I strongly endorse this volume series to the young students and promising researchers wishing to enter this exciting era of agricultural and biomedical revolutionary research. I am assured that this volume series (Essentials of Bioinformatics I–III) will provide the confidence to science students in different corners of the world, especially from the developing world with limited resources, to dream up the careers in this field to make an impact on the world.

Adeel Malik
Department of Microbiology and Molecular Biology
College of Bioscience and Biotechnology
Chungnam National University
Daejeon, Republic of Korea

Preface

Bioinformatics is an integrative field of computer science, genetics, genomics, proteomics, transcriptomics, metabolomics, and statistics, which has certainly transformed the agriculture, biology and medicine in current times. It mainly assists in modeling, predicting, and interpreting large multidimensional biological data by utilizing advanced computational methods. Currently, agricultural sciences witnessed a huge progress, where a massive data is getting generated using Omic technologies. Bioinformatics tools are essential in agricultural sciences in view of trait improvement, disease control and plant disease management, nutritional content. High-performance bioinformatics facilities in agriculture, and various bioinformatics software programs/databases are important for biotechnologists and pathologists as well as breeders.

This volume, like the first two volumes, is targeting the young researchers to make them aware of the recent developments in plant sciences, where bioinformatics along with the other multi-omics technologies changed the scientific world and is making a large impact. This volume focusses on the key development in multi-omics technologies and their impact in many aspects of plant science. Moreover, the elaboration of the database, algorithms, and software development that have been implemented to overcome the difficulties of the protein analysis are discussed.

We would also like to thank the Chairman of the Department of Biological Sciences, Prof. Khalid M. AlGhamdi, and the Head of Plant Sciences Section, Dr. Hesham F. Alharby, for providing us the valuable suggestions and encouragement to complete this task. We sincerely thank the management, faculty members, staff, and students at Princess Al-Jawhara Center of Excellence in Research of Hereditary Disorders (PACER-HD), Department of Genetic Medicine, of the Faculty of Medicine and Department of Biology of the Faculty of Science at King Abdulaziz University (KAU) for supporting our effort in bringing this book series a reality. We also appreciate the support of Prof. Jumana Y. Al-Aama, Director of PACER-HD, KAU, who, realizing the importance of bioinformatics in clinical practice, encouraged excellent scientific discussions and for supporting our work throughout this long process. We are highly grateful to all our contributors for readily accepting our invitation and for not only sharing their knowledge and research but for venerably integrating their expertise in dispersed information from diverse fields in composing the chapters and enduring editorial suggestions to finally produce this venture. We greatly appreciate their commitment. We are also thankful to Dr. Adeel Malik for his suggestions and writing the foreword for this volume. Last but not the least, we would like to acknowledge Springer Nature publishers, especially Mr. Rahul Sharma for his patience and regular communication with us to move the project forward.

Jeddah, Saudi Arabia   Khalid Rehman Hakeem
Jeddah, Saudi Arabia   Noor Ahmad Shaik
Jeddah, Saudi Arabia   Babajan Banaganapalli
Jeddah, Saudi Arabia   Ramu Elango

Contents

1 Proteoinformatics and Agricultural Biotechnology Research: Applications and Challenges 1
Jameel R. Al-Obaidi

2 Impact of Bioinformatics on Plant Science Research and Crop Improvement 29
Amrina Shafi, Insha Zahoor, Ehtishamul Haq,
and Khalid Majid Fazili

3 Bioinformatics and Plant Stress Management 47
Amrina Shafi and Insha Zahoor

4 Integration of "Omic" Approaches to Unravel the Heavy Metal Tolerance in Plants 79
Tanveer Bilal Pirzadah, Bisma Malik, and Khalid Rehman Hakeem

5 Advanced Multivariate and Computational Approaches in Agricultural Studies 93
Inayat Ur Rahman, Eduardo Soares Calixto, Aftab Afzal,
Zafar Iqbal, Niaz Ali, Farhana Ijaz, Muzammil Shah,
and Khalid Rehman Hakeem

6 Data Measurement, Data Redundancy, and Their Biological Relevance 103
Mohd Sayeed Akhtar, Ibrahim A. Alaraidh,
and Mallappa Kumara Swamy

7 Metabolomic Approaches in Plant Research 109
Ayesha T. Tahir, Qaiser Fatmi, Asia Nosheen, Mahrukh Imtiaz,
and Salma Khan

8 Bioinformatics and Medicinal Plant Research: Current Scenario ... 141
Insha Zahoor, Amrina Shafi, Khalid Majid Fazili,
and Ehtishamul Haq

9   Experimental Approaches for Genome Sequencing   159
Mohd Sayeed Akhtar, Ibrahim A. Alaraidh,
and Khalid Rehman Hakeem

10   Phylogenetic Trees: Applications, Construction, and Assessment. . . .   167
Surekha Challa and Nageswara Rao Reddy Neelapu

**11   Bioinformatics Insights on Plant Vacuolar Proton
    Pyrophosphatase: A Proton Pump Involved in Salt Tolerance**.   193
Nageswara Rao Reddy Neelapu, Sandeep Solmon Kusuma,
Titash Dutta, and Challa Surekha

Index.   213

About the Editors

Khalid Rehman Hakeem PhD, is Professor at King Abdulaziz University, Jeddah, Saudi Arabia. After completing his doctorate (Botany; specialization in Plant Ecophysiology and Molecular Biology) from Jamia Hamdard, New Delhi, India, in 2011, he worked as a Lecturer at the University of Kashmir, Srinagar, for a short period. Later, he joined Universiti Putra Malaysia, Selangor, Malaysia, and worked there as Postdoctorate Fellow in 2012 and Fellow Researcher (Associate Prof.) from 2013 to 2016. He has more than 10 years of teaching and research experience in plant ecophysiology, biotechnology and molecular biology, medicinal plant research, and plant-microbe-soil interactions as well as in environmental studies. He is the Recipient of several fellowships at both national and international levels; also, he has served as the Visiting Scientist at Jinan University, Guangzhou, China. Currently, he is involved with a number of international research projects with different government organizations.

So far, he has authored and edited more than 35 books with international publishers, including Springer Nature, Academic Press (Elsevier), and CRC Press. He also has to his credit more than 80 research publications in peer-reviewed international journals and 55 book chapters in edited volumes with international publishers.

At present, Dr. Hakeem serves as an Editorial Board Member and Reviewer of several high-impact international scientific journals from Elsevier, Springer Nature, Taylor and Francis, Cambridge, and John

Wiley Publishers. He is included in the Advisory Board of Cambridge Scholars Publishing, UK. He is also a Fellow of Plantae group of the American Society of Plant Biologists and Member of The World Academy of Sciences; the International Society for Development and Sustainability, Japan; and the Asian Federation of Biotechnology, Korea. He has been listed in Marquis Who's Who in the World, since 2014–2019. Currently, he is engaged in studying the plant processes at eco-physiological as well as molecular levels.

Noor Ahmad Shaik is an academician and researcher working in the field of Human Molecular Genetics. Over the last 15 years, he is working in the field of molecular diagnostics of different monogenic and complex disorders. With the help of high throughout molecular methods like genetic sequencing, gene expression, and metabolomics, his research team is currently trying to discover the novel causal genes/bio-markers for rare hereditary disorders. He has published 67 research publications including 54 original research articles and 13 book chapters in the fields of human genetics and bioinformatics. He has been a recipient of several research grants from different national and international funding agencies. He is currently render-ing his editorial services to world-renowned journals like Frontiers in Pediatrics and Frontiers in Genetics.

Babajan Banaganapalli (PhD) works as Bioinformatics Research Faculty at King Abdulaziz University. He initiated and successfully run the inter-disciplinary bioinformatics program from 2014 till date in King Abdulaziz University. He has more than 12 years of research experience in bioinformatics. His research interests spread across genomics, proteomics, and drug discovery for complex diseases. He has pub-lished more than 53 journal articles, conference papers, and book chapters. He has also served in numerous conference program committees, organized several bioinformatics workshops and training programs, and acted as Editor and Reviewer for various international genetics/bioinformatics journals. Recently, he was honored as Young Scientist for his outstanding research in bioinformatics by Venus International Research Foundation, India.

Ramu Elango (PhD) is a well-experienced Molecular Geneticist and Computational Biologist with extensive experience at MIT, Cambridge, USA, and GlaxoSmithKline R&D, UK, after completing his PhD in Human Genetics at All India Institute of Medical Sciences, New Delhi, India. At GlaxoSmithKline, he contributed extensively in many disease areas of interest in identifying novel causal genes and tractable drug targets. He presently heads the Research and Laboratories at the Princess Al-Jawhara Center of Excellence in Research of Hereditary Disorders, King Abdulaziz University. His research focus is on genetics and genomics of complex and polygenic diseases. His team exploits freely available large-scale genetic and genomic data with bioinformatics tools to identify the risk factors or candidate causal genes for many complex diseases.

Chapter 1
Proteoinformatics and Agricultural Biotechnology Research: Applications and Challenges

Jameel R. Al-Obaidi

Contents

1.1	Introduction	1
1.2	Proteoinformatics in Plant Disease Management	2
1.3	Proteoinformatic Databases and Tools	8
1.4	Protein-Protein Interaction Software and Database	10
1.5	Proteoinformatics of Edible Mushroom	14
1.6	Proteoinformatics of Animal Breeding Programs	14
1.7	Conclusion	15
References		16

1.1 Introduction

Bioinformatics is a multidisciplinary field incipient from the interaction of information, statistics, and biological sciences to analyze genome and/or proteome contents, sequence information, and predict the function and structure of cellular molecules that are used in construing genomics and proteomics information from an agricultural organism (Benton 1996; Bruhn and Jennings 2007). Bioinformatics is considered relatively new yet is a significant discipline within the biological sciences that has offered scientists and agrobiologists to interpret and handle huge amounts of information (Bartlett et al. 2016). This amount of data produced lead to the advancement and development of bioinformatics. The multi-"omics," together with computational biology are considered important tools in understanding genomics and its products which trigger several animal, plant, and microbial functions (Mochida and Shinozaki 2011). The functional analysis for those organisms includes profiling of gene products, prediction of interaction between proteins and their subcellular location, and also the prediction of protein metabolism pathway simulation (Xiong 2006). Bioinformatics as a tool is not isolated but frequently interacts with

J. R. Al-Obaidi (✉)
Agro-Biotechnology Institute Malaysia (ABI), National Institutes of Biotechnology
Malaysia (NIBM), 43400, Serdang, Selangor, Malaysia
e-mail: jameel@nibm.my

© Springer Nature Switzerland AG 2019
K. R. Hakeem et al. (eds.), *Essentials of Bioinformatics, Volume III*,
https://doi.org/10.1007/978-3-030-19318-8_1

other biological sections to produce assimilated results. For example, prediction of the structure of a protein depends on gene sequence and gene expression profiles, which require the use of phylogeny tools in sequence analysis. Therefore, the field of bioinformatics has developed in a way that the most important duty now include the interpretation of different types of information, including nucleotide and protein sequence and protein structures and function (Moorthie et al. 2013; Merrill et al. 2006). The analysis of DNA/RNA sequences, protein sequences and function, genome analysis and gene expression, and protein involvement in physiological functions can all make use of bioinformatic methods and tools and cannot be done without it (Collins et al. 2003). Protein sequence information and its related nucleic acid and data from many agricultural species deliver a substance for agricultural research leading to a better understanding of global agricultural needs and challenges (Kumar et al. 2015). Utilizing the available information allows and assists the identification of expression of a gene which may help to understand the relationship between phenotype and genotype (Orgogozo et al. 2015). The involvement of proteomic applications for analyses of crop, animals, and microorganisms has rapidly increased within the last decade (Mochida and Shinozaki 2010). Although proteomic approaches are regularly used in plant research worldwide, and establish powerful tools, there is still a significant area for improvement.

Proteoinformatics could be defined as "utilization of computational biology tools in the study of the proteome." Proteoinformatics is a field involving mathematics, programing sciences, statistics, and protein biology and biochemistry to predict and analyze their structure, function, and role in cell physiology (Cristoni and Mazzuca 2011; Hamady et al. 2005). Since the data obtained from agricultural proteomic research are complex and massive in size, the role of proteoinformatics is essential to reduce the time for investigation and to deliver statistically significant results and that will help to improve the plant/animal quality based on healthy growth and high productivity. Thus, proteoinformatics is a dynamic field for the development of new breed's diagnostic tools in order to develop pathogen-free/resistance and abiotic stress tolerance, high-quality traits, and higher quantity production (Koltai and Volpin 2003).

1.2   Proteoinformatics in Plant Disease Management

Among different plant pathogens, such as viruses, bacteria, and oomycetes, fungi are considered the most destructive (Dangl and Jones 2001). The growth, propagation, and survival strategies of pathogens are varied, but the strategies, in general, are similar, which start by colonization and progress to overcome host defense system and then finally infection establishment (Pegg 1981; Lawrence et al. 2016). As a result, the host-pathogen systems have led to a complex relationship between the host and the pathogen molecules, resulting in relationship with a high degree of variation (Hily et al. 2014). Proteomic studies focused mainly on the response of host plant upon pathogen attack that opened up a new era for biology in general and for agriculture in particular (Lodha et al. 2013; Alexander and Cilia 2016). Along with the use of proteomic approaches in agricultural research and the progress in sequencing agriculturally important organisms, the combination of bioinformatics and proteomics generally enhance the research in this area. This kind of multidisciplinary research is likely to fill in the gap toward the understanding of host-pathogen interaction network (Koltai and Volpin 2003). Two-dimensional gel electrophoresis has been initially used for rapidly identifying major proteome differences in control versus inoculated plants. Although many proteins identified during host-pathogen interactions have been highlighted, majority are known previously and are mainly in host immunity mechanism (Memišević et al. 2013). However, those results that arise from proteomic-based research are of great significance for the validation of gene expression in genomic or transcriptomic studies (Nesvizhskii 2014). Nevertheless, by using the gel-based proteomic tools, little novel information has been obtained, especially due to the lack of sufficient bioinformatics-related information such as genome sequences (Cho 2007). Indeed, only the most abundant proteins are detected in two-dimensional gels and successfully identified by mass spectrometry (MS). Therefore, a gap seems to be in the bioinformatics channel for the proteomics research of organisms without complete genome sequencing (Sheynkman et al. 2016). These information-related limitations in agricultural proteomic research need to be overcome to increase our knowledge on protein expression during plant-microbe interactions. However, proteomic tools have grown rapidly, and new approaches and apparatus are being developed (Mehta et al. 2008; Pérez-Clemente et al. 2013). Previous agricultural proteomics research, which mainly focused on model crops, has provided fundamental understandings into different protein families in agri-organism systems' modification and regulation (Hu et al. 2015; Vanderschuren et al. 2013). Nonetheless, model crop research itself does not retain all the information and data of interest to agricultural biology (Mirzaei et al. 2016; Carpentier et al. 2008). Therefore, those crops without complete genome sequence or sufficient genomic/EST information freely available need to be investigated (Ke et al. 2015; Ekblom and Wolf 2014). In comparison to the model organisms related to agriculture, such as rice (Koller et al. 2002), maize (Pechanova et al. 2010), chicken (Burgess 2004), cattle (Assumpcao et al. 2005), brewer's yeast (Khoa Pham and Wright 2007), and the plant pathogen *Botrytis cinerea* (Fernández-Acero et al. 2009), non-model species with little or no "bioinformation" was largely affected when it comes to proteomic analysis (Armengaud et al. 2014). Economic significance and the complexity of the genome make it necessary to sequence that organism (Bolger et al. 2014), but that is not enough to make it as a model organism if that information is not reachable by the scientific community (Canovas et al. 2004), Table 1.1 shows proteomic study of non-model organism. Most mass spectrometry proteomic methods depend on complete sequence for identification; for that reason, the analysis of these non-model species remains a challenge. Thus, relying on complete and comprehensive established database for the closely related model species "conserved genome region within the species of family" will be the only choice (Hutchins 2014; Zhu et al. 2017; Bischoff et al. 2016). However, sequence variation remains an issue, especially for quantitative proteomics

Table 1.1 Proteomic studies on non-model-pathogen interaction (2008–2018)

Plant-pathogen interaction	Proteomic platform	Main findings	References
Phaseolus vulgaris-Uromyces appendiculatus	LC-MS/MS	Resistance-genes are part of the basal system and repair disabled defenses to reinstate strong resistance	Lee et al. (2009)
cacao leaves-*Moniliophthora perniciosa*	1-DE and 2-DE	Protocols described in the study could help to develop high-level proteomic and biochemical studies in cacao also being applicable to other recalcitrant plant tissues	Pirovani et al. (2008)
*Capsicum chinense-*pepper mild mottle virus (*PMMoV*)	2-DE and MALDI-TOF/TOF	Evidence is presented for a differential accumulation of C. chinense PR proteins and mRNAs in the compatible (PMMoV-I)-C. chinense and incompatible (PMMoV-S)-C. chinense interactions for proteins belonging to all PR proteins detected	Elvira et al. (2008)
Citrus-citrus sudden death virus (CSDaV)	2-DE and MALDI-TOF/TOF	Downregulation of chitinases and proteinase inhibitors in CSD-affected plants is relevant since chitinases are well-known pathogenesis-related proteins, and their activity against plant pathogens is largely accepted	Cantú et al. (2008)
Beta vulgaris -Beet necrotic yellow vein virus (BNYVV)	MALDI-TOF-MS	Involvement of systemic resistance components in Rz1-mediated resistance and phytohormones in symptom development	Larson et al. (2008)
Glycine max-Heterodera glycines	2DE and ESI/MS-MS	Differed in resistant and susceptible Soybean Roots without cyst nematode (SCN) infestation and may form the basis of a new assay for the selection of resistance to SCN in soybean	Afzal et al. (2009)
Glycin max-Bradyrhizobium japonicum and Phytophthora sojae	2DE and quadrupole TOF MS/MS	Sap proteins from soybean that are differentially induced in response to B. japonicum and P. sojae elicitor treatments and most them were secreted proteins	Subramanian et al. (2009)
Phoenix dactylifera - Beauveria bassiana, Lecanicillium dimorphum and L. cf. psalliotae	2DE and MALDI/TOF-TOF	Proteins related with photosynthesis and energy metabolism in date palm were affected by entomopathogenic fungi colonization	Gómez-Vidal et al. (2009)

(continued)

Table 1.1 (continued)

Plant-pathogen interaction	Proteomic platform	Main findings	References
Solanum lycopersicum— Cucumber Mosaic Virus (CMV)	DIGE and MS/MS analysis	The study demonstrated that virus infection in transgenic tomato is restricted to the inoculated leaves. The study contributes to defining the role played by key proteins involved in plant-virus interaction and to studying antibody-mediated resistance	Di Carli et al. (2010)
Saccharum officinarum- Gluconacetobacter diazotrophicus	SDS-PAGE and ESI-Q-TOF	30 identified bacterial proteins in the roots of the plant samples; from those, 9 were specifically induced by plant signals	Lery et al. (2010)
Brassica juncea -Albugo candida	2DE and Q-TOF MS/MS	The study demonstrates that the timing of the expression of defense-related genes plays a crucial role during pathogenesis and incompatible interactions and that the redox balance within the chloroplast may be of crucial importance for mounting a successful defense response	Kaur et al. (2011)
Citrus aurantifolia - Candidatus Phytoplasma aurantifolia	2-DE and MS	The study provided proteomic view of the molecular basis of the infection process and identify genes that could help inhibit the effects of the pathogen	Taheri et al. (2011)
Gossypium barbadense- Verticillium dahliae	2-DE, EST database-assisted PMF and MS/MS	Infection causes elevation in ethylene biosynthesis, Bet v 1 family proteins may play an important role in the defense reaction against Verticillium wilt, and wilt resistance may implicate the redirection of carbohydrate flux from glycolysis to pentose phosphate pathway (PPP)	Wang et al. (2011)
Vitis vinifera-Uncinula necator	2-D DIGE And	High levels of Mn concentration in grapevine leaves triggered protective mechanisms against pathogens in grapevine	Yao et al. (2012)
Brassica napus -Sclerotinia sclerotiorum	2-DE and MALDI TOF/TOF	The study showed new insights into the resistance mechanisms within *B. napus* against *S. sclerotiorum*	Garg et al. (2013)
Citrus unshiu- Penicillium italicum	2-DE and LC-QToF-MS	Lignin plays important roles in heat treatment-induced citrus fruit resistance to pathogens	Yun et al. (2013)
Solanum lycopersicum - Pseudomonas syringae	iTRAQ	The study provided an insight into tomato's response to *Pseudomonas syringae*	Parker et al. (2013a)

(continued)

Table 1.1 (continued)

Plant-pathogen interaction	Proteomic platform	Main findings	References
Mentha spicata -Alternaria alternata	2-DE and MALDI TOF–TOF MSMS	The study deciphers the mechanism by which a foreign metabolite mediates stress tolerance in plant under control and infected condition	Sinha et al. (2013)
Beta vulgaris- Beet necrotic yellow vein virus	LC-MS/MS	The study identified proteins associated with systemic acquired resistance and general plant defense response	Webb et al. (2014)
Phytophthora infestans- Solanum tuberosum	LC-MS/MS	Proteins involved in sterol biosynthesis were downregulated, whereas several enzymes involved in the sesquiterpene phytoalexin biosynthesis were upregulated	Bengtsson et al. (2014)
Anacardium occidentale- Lasiodiplodia theobromae	2DE- SI-Q-TOF MS/MS	Cashew responsive proteins indicate modulation of various cellular functions involved in metabolism, stress/defense, and cell signaling	Cipriano et al. (2015)
Lactuca sativa- Salmonella enterica	2DE and nano LC-MS/MS	Proteins involved in lettuce's defense response to bacterium were upregulated, such as pyruvate dehydrogenase, 2-cys peroxiredoxin, and ferredoxin-NADP reductase	Zhang et al. (2014b)
Oil palm-Ganoderma interaction	2DE, MALDI TOF/TOF	Proteins related to lignin synthesis were downregulated up on interaction	Al-Obaidi et al. (2014)
Amorpha fruticosa- Glomus mosseae	iTRAQ and LC-MS/MS	77 proteins were classified according to different functions during the interaction	Song et al. (2015)
Vitis vinifera-Xylella fastidiosa	2DE, MALDI TOF/TOF	Muscadine and Florida hybrid bunch grapes express novel proteins in xylem to overcome pathogen attack	Katam et al. (2015b)
Solanum tuberosum - Ralstonia solanacearum	2DE and MALDI-TOF/ TOF	The study showed involvement of the identified proteins in the bacterial stress tolerance in potato	Park et al. (2016)
Malus domestica- Botryosphaeria berengeriana	2DE and MALDI-TOF- TOF	The study speculated that the upregulation of abscisic stress ripening-like protein and the dramatic decrease of -adenosylmethione synthetase in the resistant host could be related to its better disease resistance	Cai-xia et al. (2017)
Paulownia fortunei-Phytoplasma	iTRAQ	Paulownia witches' broom (PaWB) proteins may help in developing a deeper understanding of how PaWB affects the morphological characteristics of *P. fortunei*	Wei et al. (2017)

approaches, which will lead to low coverage of protein identification (Chandramouli and Qian 2009; Zhan et al. 2017). Moreover, "conserved genome" regions may produce similar protein sequence with different cellular functions and may increase the number of mismatch protein identities (Khan et al. 2014). Gel-based proteomics is considered the most dominant platform used for agricultural proteomic research (Tan et al. 2017). However, the use of gel-free proteome analysis is increasing rapidly in agricultural research with the presence of more proteoinformatics data (Porteus et al. 2011; Komatsu et al. 2013). Pathogen proteins that are used to suppress host defenses are of high importance in agricultural host-pathogen interaction, as these proteins may play a role in virulence, pathogenicity, and effector molecules (Van De Wouw and Howlett 2011). Pathogen characteristics are of primary interest in crop development programs (Fletcher et al. 2006). The contribution of proteoinformatic advances has helped the sequencing of the entire genomes of many pathogens in the last 10 years (Land et al. 2015). Classical biochemistry and molecular biology, as well as the modern omic platform techniques coupled with bioinformatic tools research, have been conducted on agricultural-related pathogens and their interactions with crops (Barah and Bones 2014). Recently, the study of pathogens have been significantly promoted by the availability of bioinformatic data and the resources for multi-"omics" research (Bhadauria 2016). These approaches, in combination with gene-targeting studies such as targeted mutations and gene silencing studies, are explained in molecular host-pathogen communications and the complex mechanisms involving pathogenesis and virulence (Allahverdiyeva et al. 2015; McGarvey et al. 2009; Fondi and Liò 2015). The present efforts to provide sufficient "proteoinformation" to determine related proteins and their function have improved the capacity to understand the core causes of crop and animal diseases and develop new possibilities of treatments (Chen et al. 2010). Proteoinformatics has many practical applications in current agricultural-related disease management with respect to the study of host-pathogen interactions, understanding the nature of the disease genetics, pathogenicity, and/or virulence factor of a pathogen which eventually aid in designing better disease control and drive the infection process which has also been identified, using molecular biological technologies and genetics in identifying the interaction with bacteria such as tomato and *Pseudomonas syringae* (Parker et al. 2013b; Balmant et al. 2015) and rice and *Xanthomonas oryzae* (Wang et al. 2013b) or with virus such as potato and potato virus (PVY) (Stare et al. 2017) or with phytopathogenic fungi such as apple and *Alternaria alternate* (Zhang et al. 2015), strawberry and *Fusarium oxysporum* (Fang et al. 2013), cotton and rot fungus *Thielaviopsis basicola* (Coumans et al. 2009), and coffee and *Hemileia vastatrix*. Proteoinformatic tools and databases related to agricultural diseases need to be further developed and expanded. Obviously, tools, software, and databases are adapted from human and more specifically medical analysis systems, and these may not necessarily be a model for analysis of crop proteomic data; therefore, more information regarding those crops and their pathogens will be very helpful to fill in the proteoinformation gap in agricultural research and also to verify the protein information predicted in the literature (Dennis et al. 2008; Thrall et al. 2011; Van Emon 2016). Generally, the proteoinformation is larger and more complicated than the genoinformation, especially in crops, since there are more proteins than genes. That is mainly because of the post-translational enzymatic modification. The nucleotide sequence can represent the genome of an organism; on the other side, peptide sequence cannot represent the proteome for that organism unless the structure of an interaction between those proteins revealed (Gupta et al. 2007; Khan 2015).

1.3 Proteoinformatic Databases and Tools

Sequencing projects of crops and animals related to agriculture bring the number of proteomic research in this field higher. Proteoinformatic methods and tools could be used to identify a specific protein of interest within the proteome of an organism which could be valuable for community related to agriculture and to interpret their cellular functions. The different and unusual protein information might be used to develop drought- and salt-tolerant crops, for diseases resistance and improvement of livestock, and higher productivity (Fears 2007; Gong et al. 2015; Ahmad et al. 2016). As discussed, a closely related sequence for a specific crop or animal can be used if genome information is not accessible. The ever-growing databases of whole genome sequence remain to accelerate capabilities of proteoinformatics, till the time of writing this chapter; there are more than 500 plants with whole genome sequence from more than 5000 eukaryotic sequence since the first genome sequence of plant (Arabidopsis) in the year 2000 (Kaul et al. 2000). Bioinformatic investigations of the genome-based information from important commercial crops revealed that gene organization over evolutionary time has remained constant and conserved, which means that knowledge obtained from model plants such as *Oryza sativa* and *Arabidopsis thaliana* may be exploited to propose food improvement programs for monocot and dicot crops, respectively (Ong et al. 2016; Jayaswal et al. 2017).

In proteoinformatics, the term "peptide/protein sequence" implies subjecting those sequences or its related databases or other methods of bioinformatics on a computer. Sequence alignment in proteoinformatics is ordering the sequences of protein/peptide, RNA, or DNA to find similar regions that may be a sign of functional and structural relationship (Pearson 2013), some important proteoinformatics databases listed in Table 1.2.

Proteoinformatics is considered as an evolving field of agricultural research. Interpreting particular functions of crops/animals is essential to determine useful proteins to improve agricultural traits (Newell-McGloughlin 2008). The integration of proteoinformatics and other omic platforms databases from agricultural species is of high importance to promote/enhance crops/animals system to solve global issues such as food, water stress, and climate changes (Katam et al. 2015a). For Asia, for instance, the Asia Pacific Bioinformatics Network (www.apbionet.org) is a good regional source (Khan et al. 2013).

Besides the classical well-known database, many website-based database or platform content have served proteomics and have been used in agricultural research.

Table 1.2 Proteoinformatics online databases/resources

Database/resources	Website link	References
Protein Information Resource	http://pir.georgetown.edu/	Wu et al. (2003)
Protein Knowledgebase	www.uniprot.org/	The UniProt Consortium (2017)
Protein domain database	http://prosite.expasy.org/	Hulo et al. (2008)
Database of Interacting Proteins	http://dip.doe-mbi.ucla.edu/dip/Main.cgi	Salwinski et al. (2004)
Large collection of protein families	http://pfam.xfam.org/	Finn et al. (2016)
Protein fingerprints	http://130.88.97.239/PRINTS/index.php	Attwood et al. (2012)
Protein data bank	http://www.wwpdb.org/	Gore et al. (2017)
Server and Repository for Protein Structure Models	https://swissmodel.expasy.org/	Biasini et al. (2014)
Database of Comparative Protein Structure Models	https://modbase.compbio.ucsf.edu/modbase-cgi/index.cgi	Pieper et al. (2014)
A General Repository for Interaction Datasets	https://thebiogrid.org/	Chatr-aryamontri et al. (2017)
Comprehensive Enzyme Information System	http://www.brenda-enzymes.org/	Placzek et al. (2017)
Encyclopedia of Genes and Genomes	http://www.kegg.jp/	Kanehisa et al. (2012)
Interacting Genes/Proteins	https://string-db.org/	Szklarczyk et al. (2017a)

The ExPASy Proteomics site, for instance, is considered as a tool developed for human proteomic research (Gasteiger et al. 2003; Hoogland et al. 2007); however, it is widely used to compute isoelectric point (pI) and molecular weight (Mw) for agricultural proteomic studies (Imam et al. 2014; Dahal et al. 2010; Guijun et al. 2006; Schneider et al. 2004; Lande et al. 2017). In general, regarding agricultural proteomics, there are a number of web-based proteomics databases that hold a plenty of efficient information (Martens 2011). Recently, a new website was developed for tracking information and articles related to the changes in plant proteomes in response to stress (PlantPReS; www.proteome.ir). Organelle proteomic analyses have also been performed in animal and plant databases such as Organelle DB (http://labs.mcdb.lsa.umich.edu/organelledb/) (Agrawal et al. 2011). Organelle expression proteomics was considered as successful tools focusing on subcellular proteins rather than total proteins (Yates Iii et al. 2005) such as mitochondrial proteome research in potato (Salvato et al. 2014), chloroplast in tomato (Tamburino et al. 2017), endoplasmic reticulum in rice (Qian et al. 2015), peroxisomes in spinach (Babujee et al. 2010), vacuoles in cauliflower (Schmidt et al. 2007), and nucleus in soybean (Cooper et al. 2011) because they have fewer proteins which can easily be identified since they contain a limited number of proteins; thus, protein identification will be more appropriate. In the last 30 years, gel-based proteomics has been used as a main platform for agricultural proteomics. The gel is stained to visualize the proteins that have travelled to specific locations in the gel. For complex samples, proteins are analyzed after enzymatic digestion (Padula et al. 2017). Many software programs were developed for gel analysis (single stained and 2D-DIGE) and used in many agricultural-related proteomic research, most of which are commercial software such as Delta2D (http://www.decodon.com/delta2d.html), ImageMaster 2D Platinum, Melanie 9 (http://2d-gel-analysis.com/), PDQuest (http://www.bio-rad.com/en-ch/product/pdquest-2-d-analysis-software), Samspots, SpotsQuest and SpotMap (http://www.cleaverscientific.com), and Dymension (http://www.syngene.com/dymension). While some of the free available software have not survived and they are either not available for download or totally discontinued such as Gel IQ from (http://ludesi.com/), there are few software which are still available and functioning (Maurer 2016; Singh 2015) such as Gel2DE, SDA for DIGE analysis, and RegStatGel (http://www.mediafire.com/FengLi/2DGelsoftware).

Followed by protein separation, the peptide MS/MS fragmented spectra are matched against the available sequence in the database for protein identification. The peptide sequence identification is obtained based on the similarity score among the experimental MS/MS and the theoretical MS/MS spectra. The mass spectra obtained during protein identification are matched with the hypothetical one existing in the database and a statistical score, based on the spectrum resemblance, is associated with the protein identification. The restraint of this approach is that only known proteins/genes reported in the database can be identified (Nilsson et al. 2010). Recently, NCBI dropped "gi number" identifier and replaced the NCBInr database with a newer database named NCBIProt which is more complicated yet more comprehensive (Disruption ahead for NCBI databases 2016). The only disadvantage of this new database is that it is time-consuming to search for non-model organism although slight improvement was noticed (data not shown). De novo sequencing can be the method of choice when the protein, in this case, the sequence is obtained directly from the MS/MS spectra to skip the step of database spectrum search. The resulted sequences are then compared with those contained in the database so to detect homologies (Ekblom and Wolf 2014).

Database search software programs/tools is listed in Table 1.3 together with those employed for de novo searching. An example of software used for de novo peptide sequencing is the Novor (www.rapidnovor.org/novor), which is capable of performing real-time de novo sequence analysis with high accuracy (Ma 2015).

1.4   Protein-Protein Interaction Software and Database

Physiological and molecular cell processes are mainly carried out through the interactions between different proteins. Interactions are physical relations between different protein structures via weak bonds (Khazanov and Carlson 2013; Chang et al. 2016). In agricultural proteomic research, identifying protein identities binding or interacting with each other during certain defined circumstances and determining

Table 1.3 List of mass spectrometry search-related software/websites

Software	Description	Website/download link	References
Maxquant	A quantitative proteomics aimed at high-resolution MS data	http://www.coxdocs.org/doku. php?id=maxquant:common:download_ and_installation	Cox et al. (2011)
Byonic™	Full MS/MS search engine providing comprehensive peptide and protein identification	https://www.proteinmetrics.com/products/ byonic/	Bern et al. (2002)
Mascot	A platform able to read various binary mass spectrometry data files	http://www.matrixscience.com/	Cottrell (2005)
MassMatrix	Database search algorithm for tandem mass spectrometric data	www.massmatrix.bio	Xu and Freitas (2009)
MS Amanda & Elutator	Scoring system to identify peptides out of tandem mass spectrometry	http://ms.imp.ac.at/i	Doblas et al. (2017)
Cyclobranch	De novo engine for identification of nonribosomal peptides	http://ms.biomed.cas.cz/cyclobranch/docs/ html/	Novák et al. (2015)
Maxquant	Quantitative proteomics software for analysis of label-free and SILAC-based proteomics	https://web.archive.org	Tyanova et al. (2016a)
SWATH	Commercial software processing tool within PeakView data can be exported for statistical analysis after false discovery rate analysis	https://sciex.com/technology/ swath-acquisition	Kang et al. (2017)

the protein-binding site are of very high importance for a better understanding of the bases of many biological/physiological activities.

Protein interactions play a significant role in protein characterization and the discovery of protein functions and the pathways they are involved in (Rao et al. 2014). This is especially true during mutualism (symbiotism), commensalism, and parasitism interaction which is caused by specific protein-protein interactions (PPI) between organisms (Leung and Poulin 2008). The precision of experimental results in revealing protein-protein interactions, however, is rather doubtful, and the availability of high-throughput platforms has shown inaccuracy and false-positive information for protein interaction. Considering experimental restrictions and limitation to find all interactions in a specific proteome, computational prediction of protein interactions is a requirement to proceed on the way to complete interactions at the proteome level (Keskin et al. 2016). Affordability of high-throughput machines and the development of computational-based prediction methods have produced vast numbers of protein-protein interactions. Computational methods for protein-protein interaction predictions can use a variety of biological data gene and protein sequences, evolution, and expression. Algorithms and statistics are commonly used to assimilate these data and deduce PPI predictions (Clark et al. 2011). This ability to provide comprehensive and reliable sets of PPIs prompted the development of many databases, aiming to gather and unify the available data, each with a different focus and different strengths. List of PPI database and examples in agriculture are presented in Table 1.4. Protein-protein interaction has been investigated and studied in many agricultural-related research such as rice with specific network (http://bis.zju.edu.cn/prin/)(Gu et al. 2011; Zhu et al. 2011), *Rhizoctonia solani*-rice interaction (Lei et al. 2014), maize (http://comp-sysbio.org/ppim/) (Zhu et al. 2015), chicken, and cattle (Fen et al. 2016).

One of the most common databases in agricultural research is the Search Tool for the Retrieval of Interacting Genes/Proteins (STRING) (Szklarczyk et al. 2015, 2017b); it is another database that incorporates both known and predicted network between proteins. Currently, STRING database covers more than 2000 species, and it is expected to cover more than 4000 in its 11th version (current version 10.5). STRING can give 3D structure besides the interaction network of a given proteome, the database used widely in prediction of protein interaction in agricultural proteomic-related research such as crop under biotic stress (Liu et al. 2015; Vu et al. 2016; Al-Obaidi et al. 2016a; Wu et al. 2015), oil-crop metabolism (Raboanatahiry et al. 2017), phytopathogenic fungi (Chu et al. 2016; Li et al. 2017), mushroom cultivation (Rahmad et al. 2014), poultry (Broiler chicken) (Zheng et al. 2016;

Table 1.4 List of protein-protein interaction (PPI) software/website

STRING	Provide a critical assessment and integration of protein-protein interactions	https://string-db.org/	Szklarczyk et al. (2017b)
MENTHA	Provides protein-protein interaction (PPI) data for many species	http://mentha.uniroma2.it/	Calderone and Cesareni (2012)
GPS-Prot	Computational prediction of phosphorylation sites	http://gps.biocuckoo.org/	Xue et al. (2008)
Compass	This tool is applicable to proteomic investigations ranging from focused studies on a small number of selected proteins	http://www.proteinsimple.com/compass/downloads/	Wenger et al. (2011)
Perseus	Shotgun proteomics data analyses	http://www.coxdocs.org/doku.php?id=perseus:start	Tyanova et al. (2016b)
Struct2Net	Structure-based protein-protein extraction	http://cb.csail.mit.edu/cb/struct2net/webserver/	Hosur et al. (2012)

Zheng et al. 2014), and buffalo (Ashok and Aparna 2017). The interactive STRING network can be recalculated based on user setting and cut-off values as well as interaction score, the maximum number of shown interactions, and expended based on user selected. Currently, it is not clear whether protein-protein interaction networks and database are representing the true biological interactomes. For that reason, agricultural proteomic researchers should depend on their own valuation of biases and consider them when inferring any knowledge based on protein interaction networks. Besides the freely available database which predict the protein-protein interaction, commercially available software platforms such as Ingenuity Pathway Analysis (IPA) (https://www.qiagenbioinformatics.com/products/ingenuity-pathway-analysis/) and Metacore (https://clarivate.com/products/metacore/) are also considered great inclusive applications that enable analysis of many "omics" (Bessarabova et al. 2012; Yin et al. 2015) and agricultural proteomics as well; however, those software applications are mainly applied in medical proteomics rather than agricultural proteomics (Chen et al. 2013).

Proteomic analysis, in general, depends on data imaging which plays a serious role in understanding new results of proteomic research. In agricultural proteomic research especially for high-throughput experiments, heat maps are particularly suitable to achieve this mission, as they allow us to find measurable forms of result presentation across proteins concurrently. It is very useful to use heat maps for presenting comparative proteomic results organized in a simple yet expressive way. The superiority of a presented heat map can be highly improved by understanding and utilizing the options available in the online tools/software to organize the data in the heat map (Key 2012; Acton 2013). The idea of a heat map style of presentation appears to be originated from the use of color-based heat maps, which used to differentiate changes in temperatures. List of used websites/software to create heat maps used in proteomic research is listed in Table 1.5.

Table 1.5 Heat map generating tools/software/website

Heatmapper	Freely available web server that allows users to interactively visualize their data in the form of heat maps	http://www.heatmapper.ca	(Babicki et al. 2016)
ComplexHeat map	Software allow users to customize heatmaps	https://bioconductor.org/packages/release/bioc/html/ComplexHeatmap.html	Gu et al. (2016)
InCHILB	Open source interactive JavaScript	https://openscreen.cz/software/inchlib/home/	Škuta et al. (2014)
InfernoRDN	Multi-omics heat map generation	https://omics.pnl.gov/software/infernordn	Sadler and Wright (2015)
Clustergrammer	Web-based tool for visualizing high-dimensional data	https://www.npmjs.com/package/clustergrammer	Fernandez et al. (2017)

1.5   Proteoinformatics of Edible Mushroom

Information regarding the life cycles and metabolisms of edible mushroom is of high importance for designing workable, fruitful, and effective cultivation process, especially with fungal species that are hard to propagate and need a special medium, temperature, etc. (Zhang et al. 2014a). Research on edible mushrooms' physiological changes, growth stages, development, interactions with the environment, and contribution in human diet used several different approaches from cell biology, physiology, and chemistry to the current and multi-omic techniques such as genomics (Chen et al. 2016), transcriptomics (Fu et al. 2017), proteomics (Rahmad et al. 2014), and metabolomics (Pandohee et al. 2015). Recently, the availability of bioinformation related many edible mushrooms species helped to conduct many proteomic researches, thanks to the availability of their genome sequencing (Shim et al. 2016; Yang et al. 2017) due to the high request for edible mushrooms and their importance in food industry, medicine, and healthcare (Yap et al. 2014).

The availability of genome sequencing for those edible mushrooms allow researchers to run genome-based proteomics (Yap et al. 2015), which provided esteemed information for initiating molecular-based markers that can be used to improve the quality and usage of edible fungi. Recently, the importance of applying proteomic platforms in edible mushroom research has been highlighted, especially with nutraceutical and medicinal application possibilities (Al-Obaidi 2016b). Mushroom genome sequences make it possible for researchers to conduct research on mushroom growth (Tang et al. 2016; Wang et al. 2013a), developmental stages (Rahmad et al. 2014; Yin et al. 2012), and higher fungi medicinal properties (Yap et al. 2014).

1.6   Proteoinformatics of Animal Breeding Programs

The final products of terrestrial (cattle, poultry, and sheep) rigorous animal agro-farming systems have conventionally been mainly meat and milk products, fish, and other products from the aquaculture segment where both gained importance in terms of capacity and nutritional properties. Fundamental proteomics can be considered a promising tool for the discovery of protein diagnostic biomarkers for different and animal product quality markers.

Recently, the interest in studying livestock animals having proteomic and metabolomic platforms have increased rapidly (Suravajhala et al. 2016). Biomarker development in chicken was identified for different research goals, while in dairy cattle, numerous potential biomarkers were detected for meat and milk production (Goldansaz et al. 2017; Ortea et al. 2016). In domestic livestock and animal proteomics, the database search identification method in general is not an issue, since a comprehensive database of protein sequences is most probably available, databases such as MetasSecKB (http://bioinformatics.ysu.edu/secretomes/animal/index.php) can be considered as a good reference. On the other side, in the cases that the animal genome has not been sequenced or not complete, other approaches such as de novo peptide sequencing is usually used (Blakeley et al. 2011). Commonly in the absence of enough proteoinformation, search against a protein sequence from closely related organisms. Small differences in peptide sequence from the sample and the genome/proteome database entries may guide to a big difference in protein identities. This issue obscures proteome analysis for non-sequenced species and between different subspecies, where the difference in the amino acid sequence of proteins is highlighted possible (Ignatchenko et al. 2017). These approaches are considered significant bioinformatic challenges because there are several aspects that affect or add inconsistency to determine protein identities. The availability of sufficient proteoinformatics data, the study of protein identification and metabolomic changes research considered the source for building models of whole systems. Such systems will permit investigators to understand the function of the protein complex in response to disease and environmental changes (Romero-Rodríguez et al. 2014). In the animal breeding proteomic research, proteomics may help in the search of animal biomarkers and offer more accurate health measures for livestock, which are essential for improving the breeding program, disease resistance, stress tolerance, and environmental changes (Marco-Ramell et al. 2016).

1.7 Conclusion

This chapter has concentrated mainly on the application of software programs and databases of proteomics in agricultural sciences, where the organism with no or incomplete genomic sequence data makes the identification of proteins more challenging in comparison to those highly studied organisms. The power of multi-omic methods for high-throughput identification and characterization of candidate genes tends to be lost in non-model organisms due to the lack of sufficient biological information. It is likely that the availability and accessibility of more sequence in plant/fungi and other agricultural-related organisms will ease some of these difficulties by making genomic data available for many non-model organisms. However, proteomic studies accumulatively produce huge amounts of data. It is usually done collecting protein annotations from databases. Answering biological questions using these data is still a great challenge. In conclusion, key objectives for agricultural proteoinformatics include the encouragement of sequence submission and make it available to the public research community. Finally, proteoinformatic databases, software programs, and methods need to be designated and utilized in a better way. Many tools and databases are adapted from human and specifically medical-related examination systems, and these may not be perfect for the analysis of plant, fungal, and other related agricultural proteomic data.

References

Acton QA (2013) Issues in bioengineering and bioinformatics: 2013 Edition. ScholarlyEditions, Atlanta

Afzal AJ, Natarajan A, Saini N, Iqbal MJ, Geisler M, El Shemy HA, Mungur R, Willmitzer L, Lightfoot DA (2009) The nematode resistance allele at the *rhg1* locus alters the proteome and primary metabolism of soybean roots. Plant Physiol 151(3):1264

Agrawal GK, Bourguignon J, Rolland N, Ephritikhine G, Ferro M, Jaquinod M, Alexiou KG, Chardot T, Chakraborty N, Jolivet P (2011) Plant organelle proteomics: collaborating for optimal cell function. Mass Spectrom Rev 30(5):772–853

Ahmad P, Latef AAA, Rasool S, Akram NA, Ashraf M, Gucel S (2016) Role of proteomics in crop stress tolerance. Front Plant Sci 7:1336

Alexander MM, Cilia M (2016) A molecular tug-of-war: global plant proteome changes during viral infection. Curr Plant Biol 5(Supplement C):13–24. https://doi.org/10.1016/j.cpb.2015.10.003

Allahverdiyeva Y, Battchikova N, Brosché M, Fujii H, Kangasjärvi S, Mulo P, Mähönen AP, Nieminen K, Overmyer K, Salojärvi J (2015) Integration of photosynthesis, development and stress as an opportunity for plant biology. New Phytol 208(3):647–655

Al-Obaidi RJ, Mohd-Yusuf Y, Razali N, Jayapalan JJ, Tey C-C, Md-Noh N, Junit MS, Othman YR, Hashim HO (2014) Identification of proteins of altered abundance in oil palm infected with Ganoderma boninense. Int J Mol Sci 15(3):5175. https://doi.org/10.3390/ijms15035175

Al-Obaidi J, Saidi N, Usuldin S, Rahmad N, Zean NB, Idris A (2016a) Differential proteomic study of oil palm leaves in response to in vitro inoculation with pathogenic and non-pathogenic Ganoderma spp. J Plant Pathol 98(2):235–244

Al-Obaidi JR (2016b) Proteomics of edible mushrooms: a mini-review. Electrophoresis 37(10):1257–1263. https://doi.org/10.1002/elps.201600031

Armengaud J, Trapp J, Pible O, Geffard O, Chaumot A, Hartmann EM (2014) Non-model organisms, a species endangered by proteogenomics. J Proteomics 105:5–18

Ashok NR, Aparna HS (2017) Empirical and bioinformatic characterization of buffalo (Bubalus bubalis) colostrum whey peptides & their angiotensin I-converting enzyme inhibition. Food Chem 228(Supplement C):582–594. https://doi.org/10.1016/j.foodchem.2017.02.032

Assumpcao TI, Fontes W, Sousa MV, Ricart CAO (2005) Proteome analysis of nelore bull (Bos taurus indicus) seminal plasma. Protein Pept Lett 12(8):813–817. https://doi.org/10.2174/0929866054864292

Attwood TK, Coletta A, Muirhead G, Pavlopoulou A, Philippou PB, Popov I, Romá-Mateo C, Theodosiou A, Mitchell AL (2012) The PRINTS database: a fine-grained protein sequence annotation and analysis resource—its status in 2012. Database 2012:bas019–bas019. https://doi.org/10.1093/database/bas019

Babicki S, Arndt D, Marcu A, Liang Y, Grant JR, Maciejewski A, Wishart DS (2016) Heatmapper: web-enabled heat mapping for all. Nucleic Acids Res 44(Web Server issue):W147–W153. https://doi.org/10.1093/nar/gkw419

Babujee L, Wurtz V, Ma C, Lueder F, Soni P, Van Dorsselaer A, Reumann S (2010) The proteome map of spinach leaf peroxisomes indicates partial compartmentalization of phylloquinone (vitamin K1) biosynthesis in plant peroxisomes. J Exp Bot 61(5):1441–1453

Balmant KM, Parker J, Yoo M-J, Zhu N, Dufresne C, Chen S (2015) Redox proteomics of tomato in response to Pseudomonas syringae infection. Hortic Res 2:15043. https://doi.org/10.1038/hortres.2015. 43. https://www.nature.com/articles/hortres201543#supplementary-information

Barah P, Bones AM (2014) Multidimensional approaches for studying plant defence against insects: from ecology to omics and synthetic biology. J Exp Bot 66(2):479–493

Bartlett A, Lewis J, Williams ML (2016) Generations of interdisciplinarity in bioinformatics. New Genet Soc 35(2):186–209. https://doi.org/10.1080/14636778.2016.1184965

Bengtsson T, Weighill D, Proux-Wéra E, Levander F, Resjö S, Burra DD, Moushib LI, Hedley PE, Liljeroth E, Jacobson D, Alexandersson E, Andreasson E (2014) Proteomics and transcriptomics of the BABA-induced resistance response in potato using a novel functional annotation approach. BMC Genomics 15(1):315. https://doi.org/10.1186/1471-2164-15-315

Benton D (1996) Bioinformatics — principles and potential of a new multidisciplinary tool. Trends Biotechnol 14(8):261–272. https://doi.org/10.1016/0167-7799(96)10037-8

Bern M, Kil YJ, Becker C (2002) Byonic: advanced peptide and protein identification software. In: Current protocols in bioinformatics. Wiley https://doi.org/10.1002/0471250953.bi1320s40

Bessarabova M, Ishkin A, JeBailey L, Nikolskaya T, Nikolsky Y (2012) Knowledge-based analysis of proteomics data. BMC Bioinforma 13(16):S13. https://doi.org/10.1186/1471-2105-13-s16-s13

Bhadauria V (2016) Omics in plant disease resistance. Caister Academic Press, Norwich

Biasini M, Bienert S, Waterhouse A, Arnold K, Studer G, Schmidt T, Kiefer F, Cassarino TG, Bertoni M, Bordoli L, Schwede T (2014) SWISS-MODEL: modelling protein tertiary and quaternary structure using evolutionary information. Nucleic Acids Res 42(W1):W252–W258. https://doi.org/10.1093/nar/gku340

Bischoff R, Permentier H, Guryev V, Horvatovich P (2016) Genomic variability and protein species—improving sequence coverage for proteogenomics. J Proteomics 134:25–36

Blakeley P, Wright JC, Hubbard SJ, Jones AR (2011) Bioinformatics in animal proteomics. In: Methods in animal proteomics, Wiley-Blackwell, pp 103–119. https://doi.org/10.1002/9780470960660.ch5

Bolger ME, Weisshaar B, Scholz U, Stein N, Usadel B, Mayer KFX (2014) Plant genome sequencing—applications for crop improvement. Curr Opin Biotechnol 26(Supplement C):31–37. https://doi.org/10.1016/j.copbio.2013.08.019

Bruhn R, Jennings SF (2007) A multidisciplinary bioinformatics minor. ACM SIGCSE Bull 39(1):348–352

Burgess S (2004) Proteomics in the chicken: tools for understanding immune responses to avian diseases. Poult Sci 83(4):552–573

Cai-xia Z, Yi T, Li-yi Z, Ze-ran Z, Pei-hua C (2017) Comparative proteomic analysis of apple branches susceptible and resistant to ring rot disease. Eur J Plant Pathol 148(2):329–341. https://doi.org/10.1007/s10658-016-1092-6

Calderone A, Cesareni G (2012) mentha: the interactome browser. 2012 18. https://doi.org/10.14806/ej.18.A.455 p. 128

Canovas FM, Dumas-Gaudot E, Recorbet G, Jorrin J, Mock HP, Rossignol M (2004) Plant proteome analysis. Proteomics 4(2):285–298

Cantú MD, Mariano AG, Palma MS, Carrilho E, Wulff NA (2008) Proteomic analysis reveals suppression of bark chitinases and proteinase inhibitors in citrus plants affected by the citrus sudden death disease. Phytopathology 98(10):1084–1092. https://doi.org/10.1094/PHYTO-98-10-1084

Carpentier SC, Panis B, Vertommen A, Swennen R, Sergeant K, Renaut J, Laukens K, Witters E, Samyn B, Devreese B (2008) Proteome analysis of non-model plants: a challenging but powerful approach. Mass Spectrom Rev 27(4):354–377

Chandramouli K, Qian P-Y (2009) Proteomics: challenges, techniques and possibilities to overcome biological sample complexity. Hum Genomics Proteomics 2009:239204. https://doi.org/10.4061/2009/239204

Chang J-W, Zhou Y-Q, Ul Qamar TM, Chen L-L, Ding Y-D (2016) Prediction of protein–protein interactions by Evidence Combining Methods. Int J Mol Sci 17(11). https://doi.org/10.3390/ijms17111946

Chatr-aryamontri A, Oughtred R, Boucher L, Rust J, Chang C, Kolas NK, O'Donnell L, Oster S, Theesfeld C, Sellam A, Stark C, Breitkreutz B-J, Dolinski K, Tyers M (2017) The BioGRID interaction database: 2017 update. Nucleic Acids Res 45(Database issue):D369–D379. https://doi.org/10.1093/nar/gkw1102

Chen C, McGarvey PB, Huang H, Wu CH (2010) Protein bioinformatics infrastructure for the integration and analysis of multiple high-throughput "omics" data. Adv Bioinforma 2010:423589. https://doi.org/10.1155/2010/423589

Chen J, Chen L, Shen B (2013) Identification of network biomarkers for cancer diagnosis. In: Wang X (ed) Bioinformatics of human proteomics. Springer Netherlands, Dordrecht, pp 257–275. https://doi.org/10.1007/978-94-007-5811-7_11

Chen L, Gong Y, Cai Y, Liu W, Zhou Y, Xiao Y, Xu Z, Liu Y, Lei X, Wang G, Guo M, Ma X, Bian Y (2016) Genome sequence of the edible cultivated mushroom lentinula edodes (Shiitake) reveals insights into lignocellulose degradation. PLoS One 11(8):e0160336. https://doi.org/10.1371/journal.pone.0160336

Cho WCS (2007) Proteomics technologies and challenges. Genomics Proteomics Bioinformatics 5(2):77–85. https://doi.org/10.1016/S1672-0229(07)60018-7

Chu X-L, Feng M-G, Ying S-H (2016) Qualitative ubiquitome unveils the potential significances of protein lysine ubiquitination in hyphal growth of Aspergillus nidulans. Curr Genet 62(1):191–201. https://doi.org/10.1007/s00294-015-0517-7

Cipriano AKAL, Gondim DMF, Vasconcelos IM, Martins JAM, Moura AA, Moreno FB, Monteiro-Moreira ACO, Melo JGM, Cardoso JE, Paiva ALS, Oliveira JTA (2015) Proteomic analysis of responsive stem proteins of resistant and susceptible cashew plants after Lasiodiplodia theobromae infection. J Proteomics 113:90–109. https://doi.org/10.1016/j.jprot.2014.09.022

Clark GW, Dar V-u-N, Bezginov A, Yang JM, Charlebois RL, Tillier ERM (2011) Using coevolution to predict protein–protein interactions. In: Cagney G, Emili A (eds) Network biology: methods and applications. Humana Press, Totowa, pp 237–256. https://doi.org/10.1007/978-1-61779-276-2_11

Collins FS, Green ED, Guttmacher AE, Guyer MS (2003) A vision for the future of genomics research. Nature 422(6934):835–847

Cooper B, Campbell KB, Feng J, Garrett WM, Frederick R (2011) Nuclear proteomic changes linked to soybean rust resistance. Mol BioSystems 7(3):773–783

Cottrell J (2005) Database searching for protein identification and characterization

Coumans JV, Poljak A, Raftery MJ, Backhouse D, Pereg-Gerk L (2009) Analysis of cotton (Gossypium hirsutum) root proteomes during a compatible interaction with the black root rot fungus Thielaviopsis basicola. Proteomics 9(2):335–349

Cox J, Neuhauser N, Michalski A, Scheltema RA, Olsen JV, Mann M (2011) Andromeda: a peptide search engine integrated into the MaxQuant environment. J Proteome Res 10(4):1794–1805. https://doi.org/10.1021/pr101065j

Cristoni S, Mazzuca S (2011) Bioinformatics applied to proteomics. In: Systems and computational biology-bioinformatics and computational modeling. IntechOpen Limited, London

Dahal D, Pich A, Braun HP, Wydra K (2010) Analysis of cell wall proteins regulated in stem of susceptible and resistant tomato species after inoculation with Ralstonia solanacearum: a proteomic approach. Plant Mol Biol 73(6):643–658. https://doi.org/10.1007/s11103-010-9646-z

Dangl JL, Jones JDG (2001) Plant pathogens and integrated defence responses to infection. Nature 411(6839):826–833

Dennis ES, Ellis J, Green A, Llewellyn D, Morell M, Tabe L, Peacock WJ (2008) Genetic contributions to agricultural sustainability. Philos Trans R Soc B: Biol Sci 363(1491):591–609. https://doi.org/10.1098/rstb.2007.2172

Di Carli M, Villani ME, Bianco L, Lombardi R, Perrotta G, Benvenuto E, Donini M (2010) Proteomic analysis of the plant–virus interaction in Cucumber Mosaic Virus (CMV) resistant transgenic tomato. J Proteome Res 9(11):5684–5697. https://doi.org/10.1021/pr100487x

Disruption ahead for NCBI databases (2016) Matrix science. http://www.matrixscience.com/blog/disruption-ahead-for-ncbi-databases.html. Accessed 20 Oct 2017

Doblas VG, Smakowska-Luzan E, Fujita S, Alassimone J, Barberon M, Madalinski M, Belkhadir Y, Geldner N (2017) Root diffusion barrier control by a vasculature-derived peptide binding to the SGN3 receptor. Science 355(6322):280

Ekblom R, Wolf JBW (2014) A field guide to whole-genome sequencing, assembly and annotation. Evol Appl 7(9):1026–1042. https://doi.org/10.1111/eva.12178

Elvira MI, Galdeano MM, Gilardi P, García-Luque I, Serra MT (2008) Proteomic analysis of pathogenesis-related proteins (PRs) induced by compatible and incompatible interactions of pepper mild mottle virus (PMMoV) in Capsicum chinense L3 plants. J Exp Bot 59(6):1253–1265. https://doi.org/10.1093/jxb/ern032

Fang X, Jost R, Finnegan PM, Barbetti MJ (2013) Comparative proteome analysis of the strawberry-fusarium oxysporum f. sp. fragariae pathosystem reveals early activation of defense responses as a crucial determinant of host resistance. J Proteome Res 12(4):1772–1788

Fears R (2007) Commission on genetic resources for food and agriculture

Fen W, Baoxing S, Xing Z, Yaotian M, Dengyun L, Na Z, Pengfei J, Qing S, Jingfei H, Deli Z (2016) Prediction and analysis of the protein-protein interaction networks for chickens, cattle, dogs, horses and rabbits. Curr Bioinform 11(1):131–142. https://doi.org/10.2174/1574893611 666151203221255

Fernandez NF, Gundersen GW, Rahman A, Grimes ML, Rikova K, Hornbeck P, Ma'ayan A (2017) Clustergrammer, a web-based heatmap visualization and analysis tool for high-dimensional biological data. Sci Data 4:170151. https://doi.org/10.1038/sdata.2017.151. https://www.nature.com/articles/sdata2017151#supplementary-information

Fernández-Acero FJ, Colby T, Harzen A, Cantoral JM, Schmidt J (2009) Proteomic analysis of the phytopathogenic fungus Botrytis cinerea during cellulose degradation. Proteomics 9(10):2892–2902

Finn RD, Coggill P, Eberhardt RY, Eddy SR, Mistry J, Mitchell AL, Potter SC, Punta M, Qureshi M, Sangrador-Vegas A, Salazar GA, Tate J, Bateman A (2016) The Pfam protein families database: towards a more sustainable future. Nucleic Acids Res 44(D1):D279–D285. https://doi.org/10.1093/nar/gkv1344

Fletcher J, Bender C, Budowle B, Cobb WT, Gold SE, Ishimaru CA, Luster D, Melcher U, Murch R, Scherm H, Seem RC, Sherwood JL, Sobral BW, Tolin SA (2006) Plant pathogen forensics: capabilities, needs, and recommendations. Microbiol Mol Biol Rev 70(2):450–471. https://doi.org/10.1128/MMBR.00022-05

Fondi M, Liò P (2015) Multi -omics and metabolic modelling pipelines: challenges and tools for systems microbiology. Microbiol Res 171(Supplement C):52–64. https://doi.org/10.1016/j.micres.2015.01.003

Fu Y, Dai Y, Yang C, Wei P, Song B, Yang Y, Sun L, Zhang Z-W, Li Y (2017) Comparative transcriptome analysis identified candidate genes related to Bailinggu mushroom formation and genetic markers for genetic analyses and breeding. Sci Rep 7:9266. https://doi.org/10.1038/s41598-017-08049-z

Garg H, Li H, Sivasithamparam K, Barbetti MJ (2013) Differentially expressed proteins and associated histological and disease progression changes in cotyledon tissue of a resistant and susceptible genotype of Brassica napus infected with Sclerotinia sclerotiorum. PLoS One 8(6):e65205. https://doi.org/10.1371/journal.pone.0065205

Gasteiger E, Gattiker A, Hoogland C, Ivanyi I, Appel RD, Bairoch A (2003) ExPASy: the proteomics server for in-depth protein knowledge and analysis. Nucleic Acids Res 31(13):3784–3788

Goldansaz SA, Guo AC, Sajed T, Steele MA, Plastow GS, Wishart DS (2017) Livestock metabolomics and the livestock metabolome: a systematic review. PLoS One 12(5):e0177675

Gómez-Vidal S, Salinas J, Tena M, Lopez-Llorca LV (2009) Proteomic analysis of date palm (Phoenix dactylifera L.) responses to endophytic colonization by entomopathogenic fungi. Electrophoresis 30(17):2996–3005. https://doi.org/10.1002/elps.200900192

Gong F, Hu X, Wang W (2015) Proteomic analysis of crop plants under abiotic stress conditions: where to focus our research? Front Plant Sci 6:418

Gore S, Sanz García E, Hendrickx PMS, Gutmanas A, Westbrook JD, Yang H, Feng Z, Baskaran K, Berrisford JM, Hudson BP, Ikegawa Y, Kobayashi N, Lawson CL, Mading S, Mak L, Mukhopadhyay A, Oldfield TJ, Patwardhan A, Peisach E, Sahni G, Sekharan MR, Sen S, Shao C, Smart OS, Ulrich EL, Yamashita R, Quesada M, Young JY, Nakamura H, Markley JL, Berman HM, Burley SK, Velankar S, Kleywegt GJ (2017) Validation of structures in the protein data bank. Structure 25(12):1916–1927. https://doi.org/10.1016/j.str.2017.10.009

Gu H, Zhu P, Jiao Y, Meng Y, Chen M (2011) PRIN: a predicted rice interactome network. BMC Bioinforma 12(1):161

Gu Z, Eils R, Schlesner M (2016) Complex heatmaps reveal patterns and correlations in multidimensional genomic data. Bioinformatics 32(18):2847–2849. https://doi.org/10.1093/bioinformatics/btw313

Guijun D, Weidong P, Gongshe L (2006) The analysis of proteome changes in sunflower seeds induced by N+ implantation. J Biosci (Bangalore) 31(2):247 253. https://doi.org/10.1007/BF02703917

Gupta N, Tanner S, Jaitly N, Adkins JN, Lipton M, Edwards R, Romine M, Osterman A, Bafna V, Smith RD, Pevzner PA (2007) Whole proteome analysis of post-translational modifications: applications of mass-spectrometry for proteogenomic annotation. Genome Res 17(9):1362–1377. https://doi.org/10.1101/gr.6427907

Hamady M, Tom Hiu Tung C, Resing K, Cios KJ, Knight R (2005) Key challenges in proteomics and proteoinformatics. IEEE Eng Med Biol Mag 24(3):34–40. https://doi.org/10.1109/MEMB.2005.1436456

Hily JM, García A, Moreno A, Plaza M, Wilkinson MD, Fereres A, Fraile A, García-Arenal F (2014) The relationship between host lifespan and pathogen reservoir potential: an analysis in the system Arabidopsis thaliana-cucumber mosaic virus. PLoS Pathog 10(11):e1004492. https://doi.org/10.1371/journal.ppat.1004492

Hoogland C, Mostaguir K, Sanchez J-C, Hochstrasser DF, Appel RD (2007) 2D PAGE databases for proteins in human body fluids. In: Thongboonkerd V (ed) Proteomics of human body fluids: principles, methods, and applications. Humana Press, Totowa, pp 137–146. https://doi.org/10.1007/978-1-59745-432-2_7

Hosur R, Peng J, Vinayagam A, Stelzl U, Xu J, Perrimon N, Bienkowska J, Berger B (2012) A computational framework for boosting confidence in high-throughput protein-protein interaction datasets. Genome Biol 13(8):R76. https://doi.org/10.1186/gb-2012-13-8-r76

Hu J, Rampitsch C, Bykova NV (2015) Advances in plant proteomics toward improvement of crop productivity and stress resistancex. Front Plant Sci 6:209. https://doi.org/10.3389/fpls.2015.00209

Hulo N, Bairoch A, Bulliard V, Cerutti L, Cuche BA, de Castro E, Lachaize C, Langendijk-Genevaux PS, Sigrist CJA (2008) The 20 years of PROSITE. Nucleic Acids Res 36(Database issue):D245–D249. https://doi.org/10.1093/nar/gkm977

Hutchins JRA (2014) What's that gene (or protein)? Online resources for exploring functions of genes, transcripts, and proteins. Mol Biol Cell 25(8):1187–1201. https://doi.org/10.1091/mbc.E13-10-0602

Ignatchenko A, Sinha A, Alfaro JA, Boutros PC, Kislinger T, Ignatchenko V (2017) Detecting protein variants by mass spectrometry: a comprehensive study in cancer cell-lines. Genome Med 9(1):62

Imam J, Nitin M, Toppo NN, Mandal NP, Kumar Y, Variar M, Bandopadhyay R, Shukla P (2014) A comprehensive overview on application of bioinformatics and computational statistics in rice genomics toward an Amalgamated approach for improving acquaintance base. In: Kishor PBK, Bandopadhyay R, Suravajhala P (eds) Agricultural bioinformatics. Springer India, New Delhi, pp 89–107. https://doi.org/10.1007/978-81-322-1880-7_5

Jayaswal PK, Dogra V, Shanker A, Sharma TR, Singh NK (2017) A tree of life based on ninety-eight expressed genes conserved across diverse eukaryotic species. PLoS One 12(9):e0184276. https://doi.org/10.1371/journal.pone.0184276

Kanehisa M, Goto S, Sato Y, Furumichi M, Tanabe M (2012) KEGG for integration and interpretation of large-scale molecular data sets. Nucleic Acids Res 40(D1):D109–D114. https://doi.org/10.1093/nar/gkr988

Kang Y, Burton L, Lau A, Tate S (2017) SWATH-ID: an instrument method which combines identification and quantification in a single analysis. Proteomics 17(10):1500522-n/a. https://doi.org/10.1002/pmic.201500522

Katam K, Jones KA, Sakata K (2015a) Advances in proteomics and bioinformatics in agriculture research and crop improvement. J Proteomics Bioinforma 8(3):39

Katam R, Chibanguza K, Latinwo LM, Smith D (2015b) Proteome biomarkers in xylem reveal Pierce's disease tolerance in grape. J Proteomics Bioinforma 8(9):217–224

Kaul S, Koo HL, Jenkins J, Rizzo M, Rooney T, Tallon LJ, Feldblyum T, Nierman W, Benito MI, Lin X (2000) Analysis of the genome sequence of the flowering plant Arabidopsis thaliana. Nature 408(6814):796–815

Kaur P, Jost R, Sivasithamparam K, Barbetti MJ (2011) Proteome analysis of the Albugo candida–Brassica juncea pathosystem reveals that the timing of the expression of defence-related genes is a crucial determinant of pathogenesis. J Exp Bot 62(3):1285–1298. https://doi.org/10.1093/jxb/erq365

Ke T, Yu J, Dong C, Mao H, Hua W, Liu S (2015) ocsESTdb: a database of oil crop seed EST sequences for comparative analysis and investigation of a global metabolic network and oil accumulation metabolism. BMC Plant Biol 15(1):19. https://doi.org/10.1186/s12870-014-0399-8

Keskin O, Tuncbag N, Gursoy A (2016) Predicting protein–protein interactions from the molecular to the proteome level. Chem Rev 116(8):4884–4909

Key M (2012) A tutorial in displaying mass spectrometry-based proteomic data using heat maps. BMC Bioinform 13(Suppl 16):S10–S10. https://doi.org/10.1186/1471-2105-13-S16-S10

Khan FA (2015) Biotechnology fundamentals, 2nd edn. CRC Press, Boca Raton, Florida, United States

Khan AM, Tan TW, Schönbach C, Ranganathan S (2013) APBioNet—Transforming Bioinformatics in the Asia-Pacific Region. PLoS Comp Biol 9(10):e1003317. https://doi.org/10.1371/journal.pcbi.1003317

Khan I, Chen Y, Dong T, Hong X, Takeuchi R, Mori H, Kihara D (2014) Genome-scale identification and characterization of moonlighting proteins. Biol Direct 9(1):30. https://doi.org/10.1186/s13062-014-0030-9

Khazanov NA, Carlson HA (2013) Exploring the composition of protein-ligand binding sites on a large scale. PLoS Comp Biol 9(11):e1003321. https://doi.org/10.1371/journal.pcbi.1003321

Khoa Pham T, Wright PC (2007) Proteomic analysis of Saccharomyces cerevisiae. Expert Rev Proteomics 4(6):793–813

Koller A, Washburn MP, Lange BM, Andon NL, Deciu C, Haynes PA, Hays L, Schieltz D, Ulaszek R, Wei J, Wolters D, Yates JR (2002) Proteomic survey of metabolic pathways in rice. Proc Natl Acad Sci 99(18):11969–11974

Koltai H, Volpin H (2003) Agricultural genomics: an approach to plant protection. Eur J Plant Pathol 109(2):101–108. https://doi.org/10.1023/A:1022512914003

Komatsu S, Mock H-P, Yang P, Svensson B (2013) Application of proteomics for improving crop protection/artificial regulation. Front Plant Sci 4:522. https://doi.org/10.3389/fpls.2013.00522

Kumar A, Pathak RK, Gupta SM, Gaur VS, Pandey D (2015) Systems biology for smart crops and agricultural innovation: filling the gaps between genotype and phenotype for complex traits linked with robust agricultural productivity and sustainability. OMICS: J Integr Biol 19(10):581–601

Land M, Hauser L, Jun S-R, Nookaew I, Leuze MR, Ahn T-H, Karpinets T, Lund O, Kora G, Wassenaar T, Poudel S, Ussery DW (2015) Insights from 20 years of bacterial genome sequencing. Funct Integr Genomics 15(2):141–161. https://doi.org/10.1007/s10142-015-0433-4

Lande NV, Subba P, Barua P, Gayen D, Keshava Prasad TS, Chakraborty S, Chakraborty N (2017) Dissecting the chloroplast proteome of chickpea (Cicer arietinum L.) provides new insights into classical and non-classical functions. J Proteomics 165(Supplement C):11–20. https://doi.org/10.1016/j.jprot.2017.06.005

Larson RL, Wintermantel WM, Hill A, Fortis L, Nunez A (2008) Proteome changes in sugar beet in response to Beet necrotic yellow vein virus. Physiol Mol Plant Pathol 72(1):62–72. https://doi.org/10.1016/j.pmpp.2008.04.003

Lawrence S, Parker J, Chen S (2016) Plant response to bacterial pathogens: a proteomics view. In: Salekdeh GH (ed) Agricultural proteomics volume 2: environmental stresses. Springer International Publishing, Cham, pp 203–225. https://doi.org/10.1007/978-3-319-43278-6_9

Lee J, Feng J, Campbell KB, Scheffler BE, Garrett WM, Thibivilliers S, Stacey G, Naiman DQ, Tucker ML, Pastor-Corrales MA, Cooper B (2009) Quantitative proteomic analysis of bean plants infected by a virulent and avirulent obligate rust fungus. Mol Cell Proteomics 8(1):19–31. https://doi.org/10.1074/mcp.M800156-MCP200

Lei D, Lin R, Yin C, Li P, Zheng A (2014) Global protein–protein interaction network of rice sheath blight pathogen. J Proteome Res 13(7):3277–3293. https://doi.org/10.1021/pr500069r

Lery LMS, Hemerly AS, Nogueira EM, von Krüger WMA, Bisch PM (2010) Quantitative proteomic analysis of the interaction between the endophytic plant-growth-promoting bacterium gluconacetobacter diazotrophicus and sugarcane. Mol Plant Microbe Interact 24(5):562–576. https://doi.org/10.1094/MPMI-08-10-0178

Leung T, Poulin R (2008) Parasitism, commensalism, and mutualism: exploring the many shades of symbioses. Vie Milieu 58(2):107–115

Li T, Gong L, Wang Y, Chen F, Gupta VK, Jian Q, Duan X, Jiang Y (2017) Proteomics analysis of Fusarium proliferatum under various initial pH during fumonisin production. J Proteomics 164(Supplement C):59–72. https://doi.org/10.1016/j.jprot.2017.05.008

Liu W, Gray S, Huo Y, Li L, Wei T, Wang X (2015) Proteomic analysis of interaction between a plant virus and its vector insect reveals new functions of hemipteran cuticular protein. Mol Cell Proteomics: MCP 14(8):2229–2242. https://doi.org/10.1074/mcp.M114.046763

Lodha TD, Hembram P, Nitile Tep JB (2013) Proteomics: a successful approach to understand the molecular mechanism of plant-pathogen interaction. Am J Plant Sci 04(06):15. https://doi.org/10.4236/ajps.2013.46149

Ma B (2015) Novor: real-time peptide de novo sequencing software. J Am Soc Mass Spectrom 26(11):1885–1894. https://doi.org/10.1007/s13361-015-1204-0

Marco-Ramell A, de Almeida AM, Cristobal S, Rodrigues P, Roncada P, Bassols A (2016) Proteomics and the search for welfare and stress biomarkers in animal production in the one-health context. Mol Biosyst 12(7):2024–2035

Martens L (2011) Proteomics databases and repositories. In: Wu CH, Chen C (eds) Bioinformatics for comparative proteomics. Humana Press, Totowa, pp 213–227. https://doi.org/10.1007/978-1-60761-977-2_14

Maurer MH (2016) Two-dimensional gel electrophoresis image analysis via dedicated software packages. In: Marengo E, Robotti E (eds) 2-D PAGE map analysis: methods and protocols. Springer New York, New York, pp 55–65. https://doi.org/10.1007/978-1-4939-3255-9_3

McGarvey PB, Huang H, Mazumder R, Zhang J, Chen Y, Zhang C, Cammer S, Will R, Odle M, Sobral B, Moore M, Wu CH (2009) Systems integration of biodefense omics data for analysis of pathogen-host interactions and identification of potential targets. PLoS One 4(9):e7162. https://doi.org/10.1371/journal.pone.0007162

Mehta A, Brasileiro ACM, Souza DSL, Romano E, Campos MA, Grossi-de-Sá MF, Silva MS, Franco OL, Fragoso RR, Bevitori R, Rocha TL (2008) Plant–pathogen interactions: what is proteomics telling us? FEBS J 275(15):3731–3746. https://doi.org/10.1111/j.1742-4658.2008.06528.x

Memišević V, Zavaljevski N, Pieper R, Rajagopala SV, Kwon K, Townsend K, Yu C, Yu X, DeShazer D, Reifman J, Wallqvist A (2013) Novel Burkholderia mallei virulence factors linked to specific host-pathogen protein interactions. Mol Cell Proteomics : MCP 12(11):3036–3051. https://doi.org/10.1074/mcp.M113.029041

Merrill SA, Mazza A-M, Council NR (2006) Genomics, proteomics, and the changing research environment

Mirzaei M, Wu Y, Handler D, Maher T, Pascovici D, Ravishankar P, Moghaddam MZ, Haynes PA, Salekdeh GH, Chick JM (2016) Applications of quantitative proteomics in plant research. In: Agricultural proteomics volume 1. Springer, Switzerland. pp 1–29

Mochida K, Shinozaki K (2010) Genomics and bioinformatics resources for crop improvement. Plant Cell Physiol 51(4):497–523. https://doi.org/10.1093/pcp/pcq027

Mochida K, Shinozaki K (2011) Advances in omics and bioinformatics tools for systems analyses of plant functions. Plant Cell Physiol 52(12):2017–2038. https://doi.org/10.1093/pcp/pcr153

Moorthie S, Hall A, Wright CF (2013) Informatics and clinical genome sequencing: opening the black box. Genet Med 15(3):165–171

Nesvizhskii AI (2014) Proteogenomics: concepts, applications, and computational strategies. Nat Methods 11(11):1114–1125. https://doi.org/10.1038/nmeth.3144

Newell-McGloughlin M (2008) Nutritionally improved agricultural crops. Plant Physiol 147(3):939–953. https://doi.org/10.1104/pp.108.121947

Nilsson T, Mann M, Aebersold R, Yates JR, Bairoch A, Bergeron JJM (2010) Mass spectrometry in high-throughput proteomics: ready for the big time. Nat Methods 7(9):681–685

Novák J, Lemr K, Schug KA, Havlíček V (2015) CycloBranch: de novo sequencing of non-ribosomal peptides from accurate product ion mass spectra. J Am Soc Mass Spectrom 26(10):1780–1786. https://doi.org/10.1007/s13361-015-1211-1

Ong Q, Nguyen P, Phuong Thao N, Le L (2016) Bioinformatics approach in plant genomic research. Curr Genomics 17(4):368–378

Orgogozo V, Morizot B, Martin A (2015) The differential view of genotype–phenotype relationships. Front Genet 6:179. https://doi.org/10.3389/fgene.2015.00179

Ortea I, O'Connor G, Maquet A (2016) Review on proteomics for food authentication. J Proteomics 147:212–225

Padula PM, Berry JI, O'Rourke BM, Raymond BB, Santos J, Djordjevic SP (2017) A comprehensive guide for performing sample preparation and top-down protein analysis. Proteomes 5(2):11. https://doi.org/10.3390/proteomes5020011

Pandohee J, Stevenson PG, Conlan XA, Zhou X-R, Jones OAH (2015) Off-line two-dimensional liquid chromatography for metabolomics: an example using Agaricus bisporus mushrooms exposed to UV irradiation. Metabolomics 11 (4):939–951. https://doi.org/10.1007/s11306-014-0749-4

Park S, Gupta R, Krishna R, Kim ST, Lee DY, Hwang D-j, Bae S-C, Ahn I-P (2016) Proteome analysis of disease resistance against Ralstonia solanacearum in Potato Cultivar CT206-10. Plant Pathol J 32(1):25–32. https://doi.org/10.5423/PPJ.OA.05.2015.0076

Parker J, Koh J, Yoo M-J, Zhu N, Feole M, Yi S, Chen S (2013a) Quantitative proteomics of tomato defense against Pseudomonas syringae infection. Proteomics 13(12–13):1934–1946. https://doi.org/10.1002/pmic.201200402

Parker J, Koh J, Yoo MJ, Zhu N, Feole M, Yi S, Chen S (2013b) Quantitative proteomics of tomato defense against Pseudomonas syringae infection. Proteomics 13(12–13):1934–1946

Pearson WR (2013) An introduction to sequence similarity ("Homology") searching. Curr Protoc Bioinform/editoral board, Andreas D Baxevanis [et al] 0 3:10.1002/0471250953.bi0471250301s0471250942. https://doi.org/10.1002/0471250953.bi0301s42

Pechanova O, Pechan T, Ozkan S, McCarthy FM, Williams WP, Luthe DS (2010) Proteome profile of the developing maize (Zea mays L.) rachis. Proteomics 10(16):3051–3055. https://doi.org/10.1002/pmic.200900833

Pegg GF (1981) Chapter 7 – Biochemistry and physiology of pathogenesis. In: Fungal Wilt diseases of plants, Academic Press, Cambridge, Massachusetts, United States. pp 193–253. https://doi.org/10.1016/B978-0-12-464450-2.50012-7

Pérez-Clemente RM, Vives V, Zandalinas SI, López-Climent MF, Muñoz V, Gómez-Cadenas A (2013) Biotechnological approaches to study plant responses to stress. Biomed Res Int 2013:654120. https://doi.org/10.1155/2013/654120

Pieper U, Webb BM, Dong GQ, Schneidman-Duhovny D, Fan H, Kim SJ, Khuri N, Spill YG, Weinkam P, Hammel M, Tainer JA, Nilges M, Sali A (2014) ModBase, a database of annotated comparative protein structure models and associated resources. Nucleic Acids Res 42(D1):D336–D346. https://doi.org/10.1093/nar/gkt1144

Pirovani CP, Carvalho HAS, Machado RCR, Gomes DS, Alvim FC, Pomella AWV, Gramacho KP, Cascardo JCM, Pereira GAG, Micheli F (2008) Protein extraction for proteome analysis from cacao leaves and meristems, organs infected by Moniliophthora perniciosa, the causal agent of the witches' broom disease. Electrophoresis 29(11):2391–2401. https://doi.org/10.1002/elps.200700743

Placzek S, Schomburg I, Chang A, Jeske L, Ulbrich M, Tillack J, Schomburg D (2017) BRENDA in 2017: new perspectives and new tools in BRENDA. Nucleic Acids Res 45(D1):D380–D388. https://doi.org/10.1093/nar/gkw952

Porteus B, Kocharunchitt C, Nilsson RE, Ross T, Bowman JP (2011) Utility of gel-free, label-free shotgun proteomics approaches to investigate microorganisms. Appl Microbiol Biotechnol 90(2):407–416. https://doi.org/10.1007/s00253-011-3172-z

Qian D, Tian L, Qu L (2015) Proteomic analysis of endoplasmic reticulum stress responses in rice seeds. Sci Rep 5:14255. https://doi.org/10.1038/srep14255. https://www.nature.com/articles/srep14255#supplementary-information

Raboanatahiry N, Chao H, Guo L, Gan J, Xiang J, Yan M, Zhang L, Yu L, Li M (2017) Synteny analysis of genes and distribution of loci controlling oil content and fatty acid profile based on QTL alignment map in Brassica napus. BMC Genomics 18(1):776. https://doi.org/10.1186/s12864-017-4176-6

Rahmad N, Al-Obaidi JR, Rashid NMN, Zean NB, Yusoff MHYM, Shaharuddin NS, Jamil NAM, Saleh NM (2014) Comparative proteomic analysis of different developmental stages of the edible mushroom Termitomyces heimii. Biol Res 47(1):30. https://doi.org/10.1186/0717-6287-47-30

Rao VS, Srinivas K, Sujini GN, Kumar GNS (2014) Protein-protein interaction detection: methods and analysis. Int J Proteomics 2014:12. https://doi.org/10.1155/2014/147648

Romero-Rodríguez MC, Pascual J, Valledor L, Jorrín-Novo J (2014) Improving the quality of protein identification in non-model species. Characterization of Quercus ilex seed and Pinus radiata needle proteomes by using SEQUEST and custom databases. J Proteomics 105:85–91

Sadler NC, Wright AT (2015) Activity-based protein profiling of microbes. Curr Opin Chem Biol 24:139–144. https://doi.org/10.1016/j.cbpa.2014.10.022

Salvato F, Havelund JF, Chen M, Rao RSP, Rogowska-Wrzesinska A, Jensen ON, Gang DR, Thelen JJ, Møller IM (2014) The potato tuber mitochondrial proteome. Plant Physiol 164(2):637

Salwinski L, Miller CS, Smith AJ, Pettit FK, Bowie JU, Eisenberg D (2004) The database of interacting proteins: 2004 update. Nucleic Acids Res 32(Suppl_1):D449–D451. https://doi.org/10.1093/nar/gkh086

Schmidt UG, Endler A, Schelbert S, Brunner A, Schnell M, Neuhaus HE, Marty-Mazars D, Marty F, Baginsky S, Martinoia E (2007) Novel tonoplast transporters identified using a proteomic approach with vacuoles isolated from cauliflower buds. Plant Physiol 145(1):216

Schneider M, Tognolli M, Bairoch A (2004) The Swiss-Prot protein knowledgebase and ExPASy: providing the plant community with high quality proteomic data and tools. Plant Physiol Biochem 42(12):1013–1021. https://doi.org/10.1016/j.plaphy.2004.10.009

Sheynkman GM, Shortreed MR, Cesnik AJ, Smith LM (2016) Proteogenomics: integrating next-generation sequencing and mass spectrometry to characterize human proteomic variation. Annu Rev Anal Chem (Palo Alto, Calif) 9(1):521–545. https://doi.org/10.1146/annurev-anchem-071015-041722

Shim D, Park S-G, Kim K, Bae W, Lee GW, Ha B-S, Ro H-S, Kim M, Ryoo R, Rhee S-K, Nou I-S, Koo C-D, Hong CP, Ryu H (2016) Whole genome de novo sequencing and genome annotation of the world popular cultivated edible mushroom, Lentinula edodes. J Biotechnol 223(Supplement C):24–25. https://doi.org/10.1016/j.jbiotec.2016.02.032

Singh R (2015) Bioinformatics: genomics and proteomics. Vikas Publishing House, Chennai, India

Sinha R, Bhattacharyya D, Majumdar AB, Datta R, Hazra S, Chattopadhyay S (2013) Leaf proteome profiling of transgenic mint infected with Alternaria alternata. J Proteomics 93:117–132. https://doi.org/10.1016/j.jprot.2013.01.020

Škuta C, Bartůněk P, Svozil D (2014) InCHlib – interactive cluster heatmap for web applications. J Cheminformatics 6(1):44. https://doi.org/10.1186/s13321-014-0044-4

Song F, Qi D, Liu X, Kong X, Gao Y, Zhou Z, Wu Q (2015) Proteomic analysis of symbiotic proteins of Glomus mosseae and Amorpha fruticosa. Sci Rep 5:18031. https://doi.org/10.1038/srep18031. https://www.nature.com/articles/srep18031#supplementary-information

Stare T, Stare K, Weckwerth W, Wienkoop S, Gruden K (2017) Comparison between proteome and transcriptome response in potato (Solanum tuberosum L.) leaves following Potato virus Y (PVY) infection. Proteomes 5(3):14

Subramanian S, Cho U-H, Keyes C, Yu O (2009) Distinct changes in soybean xylem sap proteome in response to pathogenic and symbiotic microbe interactions. BMC Plant Biol 9(1):119. https://doi.org/10.1186/1471-2229-9-119

Suravajhala P, Kogelman LJ, Kadarmideen HN (2016) Multi-omic data integration and analysis using systems genomics approaches: methods and applications in animal production, health and welfare. Genet Sel Evol 48(1):38

Szklarczyk D, Franceschini A, Wyder S, Forslund K, Heller D, Huerta-Cepas J, Simonovic M, Roth A, Santos A, Tsafou KP, Kuhn M, Bork P, Jensen LJ, von Mering C (2015) STRING v10: protein–protein interaction networks, integrated over the tree of life. Nucleic Acids Res 43(Database issue):D447–D452. https://doi.org/10.1093/nar/gku1003

Szklarczyk D, Morris JH, Cook H, Kuhn M, Wyder S, Simonovic M, Santos A, Doncheva NT, Roth A, Bork P (2017a) The STRING database in 2017: quality-controlled protein–protein association networks, made broadly accessible. Nucleic Acids Res 45(D1):D362–D368

Szklarczyk D, Morris JH, Cook H, Kuhn M, Wyder S, Simonovic M, Santos A, Doncheva NT, Roth A, Bork P, Jensen LJ, von Mering C (2017b) The STRING database in 2017: quality-controlled protein–protein association networks, made broadly accessible. Nucleic Acids Res 45(Database issue):D362–D368. https://doi.org/10.1093/nar/gkw937

Taheri F, Nematzadeh G, Zamharir MG, Nekouei MK, Naghavi M, Mardi M, Salekdeh GH (2011) Proteomic analysis of the Mexican lime tree response to "Candidatus Phytoplasma aurantifolia" infection. Mol BioSystems 7(11):3028–3035. https://doi.org/10.1039/c1mb05268c

Tamburino R, Vitale M, Ruggiero A, Sassi M, Sannino L, Arena S, Costa A, Batelli G, Zambrano N, Scaloni A, Grillo S, Scotti N (2017) Chloroplast proteome response to drought stress and recovery in tomato (Solanum lycopersicum L.). BMC Plant Biol 17(1):40. https://doi.org/10.1186/s12870-017-0971-0

Tan BC, Lim YS, Lau S-E (2017) Proteomics in commercial crops: an overview. J Proteomics. https://doi.org/10.1016/j.jprot.2017.05.018

Tang LH, Tan Q, Bao DP, Zhang XH, Jian HH, Li Y, Yang R, Wang Y (2016) Comparative proteomic analysis of light-induced mycelial Brown film formation in Lentinula edodes. Biomed Res Int 2016:8. https://doi.org/10.1155/2016/5837293

The UniProt Consortium (2017) UniProt: the universal protein knowledgebase. Nucleic Acids Res 45(Database issue):D158–D169. https://doi.org/10.1093/nar/gkw1099

Thrall PH, Oakeshott JG, Fitt G, Southerton S, Burdon JJ, Sheppard A, Russell RJ, Zalucki M, Heino M, Ford Denison R (2011) Evolution in agriculture: the application of evolutionary approaches to the management of biotic interactions in agro-ecosystems. Evol Appl 4(2):200–215. https://doi.org/10.1111/j.1752-4571.2010.00179.x

Tyanova S, Temu T, Cox J (2016a) The MaxQuant computational platform for mass spectrometry-based shotgun proteomics. Nat Protoc 11:2301. https://doi.org/10.1038/nprot.2016.136

Tyanova S, Temu T, Sinitcyn P, Carlson A, Hein MY, Geiger T, Mann M, Cox J (2016b) The Perseus computational platform for comprehensive analysis of (prote)omics data. Nat Methods 13:731. https://doi.org/10.1038/nmeth.3901. https://www.nature.com/articles/nmeth.3901#supplementary-information

Van De Wouw AP, Howlett BJ (2011) Fungal pathogenicity genes in the age of 'omics'. Mol Plant Pathol 12(5):507–514

Van Emon JM (2016) The omics revolution in agricultural research. J Agric Food Chem 64(1):36–44. https://doi.org/10.1021/acs.jafc.5b04515

Vanderschuren H, Lentz E, Zainuddin I, Gruissem W (2013) Proteomics of model and crop plant species: status, current limitations and strategic advances for crop improvement. J Proteomics 93:5–19

Vu LD, Stes E, Van Bel M, Nelissen H, Maddelein D, Inzé D, Coppens F, Martens L, Gevaert K, De Smet I (2016) Up-to-date workflow for plant (Phospho)proteomics identifies differential drought-responsive phosphorylation events in Maize leaves. J Proteome Res 15(12):4304–4317. https://doi.org/10.1021/acs.jproteome.6b00348

Wang F-X, Ma Y-P, Yang C-L, Zhao P-M, Yao Y, Jian G-L, Luo Y-M, Xia G-X (2011) Proteomic analysis of the sea-island cotton roots infected by wilt pathogen Verticillium dahliae. Proteomics 11(22):4296–4309. https://doi.org/10.1002/pmic.201100062

Wang M, Gu B, Huang J, Jiang S, Chen Y, Yin Y, Pan Y, Yu G, Li Y, Wong BHC, Liang Y, Sun H (2013a) Transcriptome and proteome exploration to provide a resource for the study of Agrocybe aegerita. PLoS One 8(2):e56686. https://doi.org/10.1371/journal.pone.0056686

Wang Y, Kim SG, Wu J, Huh HH, Lee SJ, Rakwal R, Kumar Agrawal G, Park ZY, Young Kang K, Kim ST (2013b) Secretome analysis of the rice bacterium Xanthomonas oryzae (Xoo) using in vitro and in planta systems. Proteomics 13(12–13):1901–1912

Webb K, Broccardo C, Prenni J, Wintermantel W (2014) Proteomic profiling of sugar beet (Beta vulgaris) leaves during rhizomania compatible interactions. Proteomes 2(2):208

Wei Z, Wang Z, Li X, Zhao Z, Deng M, Dong Y, Cao X, Fan G (2017) Comparative proteomic analysis of Paulownia fortunei response to phytoplasma infection with dimethyl sulfate treatment. Int J Genomics 2017:11. https://doi.org/10.1155/2017/6542075

Wenger CD, Phanstiel DH, Lee MV, Bailey DJ, Coon JJ (2011) COMPASS: a suite of pre- and post-search proteomics software tools for OMSSA. Proteomics 11(6):1064–1074. https://doi.org/10.1002/pmic.201000616

Wu CH, Yeh L-SL, Huang H, Arminski L, Castro-Alvear J, Chen Y, Hu Z, Kourtesis P, Ledley RS, Suzek BE (2003) The protein information resource. Nucleic Acids Res 31(1):345–347

Wu L, Wang S, Wu J, Han Z, Wang R, Wu L, Zhang H, Chen Y, Hu X (2015) Phosphoproteomic analysis of the resistant and susceptible genotypes of maize infected with sugarcane mosaic virus. Amino Acids 47(3):483–496. https://doi.org/10.1007/s00726-014-1880-2

Xiong J (2006) Essential bioinformatics. Cambridge University Press, United Kingdom

Xu H, Freitas MA (2009) MassMatrix: a database search program for rapid characterization of proteins and peptides from Tandem Mass Spectrometry Data. Proteomics 9(6):1548–1555. https://doi.org/10.1002/pmic.200700322

Xue Y, Ren J, Gao X, Jin C, Wen L, Yao X (2008) GPS 2.0, a tool to predict kinase-specific phosphorylation sites in hierarchy. Mol Cell Proteomics: MCP 7(9):1598–1608. https://doi.org/10.1074/mcp.M700574-MCP200

Yang R, Li Y, Song X, Tang L, Li C, Tan Q, Bao D (2017) The complete mitochondrial genome of the widely cultivated edible fungus Lentinula edodes. Mitochondrial DNA Part B 2(1):13–14. https://doi.org/10.1080/23802359.2016.1275839

Yao YA, Wang J, Ma X, Lutts S, Sun C, Ma J, Yang Y, Achal V, Xu G (2012) Proteomic analysis of Mn-induced resistance to powdery mildew in grapevine. J Exp Bot 63(14):5155–5170. https://doi.org/10.1093/jxb/ers175

Yap H-YY, Chooi Y-H, Firdaus-Raih M, Fung S-Y, Ng S-T, Tan C-S, Tan N-H (2014) The genome of the Tiger Milk mushroom, Lignosus rhinocerotis, provides insights into the genetic basis of its medicinal properties. BMC Genomics 15(1):635. https://doi.org/10.1186/1471-2164-15-635

Yap H-YY, Fung S-Y, Ng S-T, Tan C-S, Tan N-H (2015) Genome-based proteomic analysis of Lignosus rhinocerotis (Cooke) Ryvarden sclerotium. Int J Med Sci 12(1):23

Yates Iii JR, Gilchrist A, Howell KE, Bergeron JJM (2005) Proteomics of organelles and large cellular structures. Nat Rev Mol Cell Biol 6:702. https://doi.org/10.1038/nrm1711

Yin Y, Yu G, Chen Y, Jiang S, Wang M, Jin Y, Lan X, Liang Y, Sun H (2012) Genome-wide transcriptome and proteome analysis on different developmental stages of Cordyceps militaris. PLoS One 7(12):e51853

Yin S-Y, Pradeep MS, Yang N-S (2015) Use of omics approaches for developing immune-modulatory and anti-inflammatory phytomedicines. In: Genomics, proteomics and metabolomics in nutraceuticals and functional foods. Wiley, pp 453–475. https://doi.org/10.1002/9781118930458.ch36

Yun Z, Gao H, Liu P, Liu S, Luo T, Jin S, Xu Q, Xu J, Cheng Y, Deng X (2013) Comparative proteomic and metabolomic profiling of citrus fruit with enhancement of disease resistance by postharvest heat treatment. BMC Plant Biol 13(1):44. https://doi.org/10.1186/1471-2229-13-44

Zhan X, Long Y, Lu M (2017) Exploration of variations in proteome and metabolome for predictive diagnostics and personalized treatment algorithms: Innovative approach and examples for potential clinical application. J Proteomics. https://doi.org/10.1016/j.jprot.2017.08.020

Zhang Y, Geng W, Shen Y, Wang Y, Dai Y-C (2014a) Edible mushroom cultivation for food security and rural development in China: bio-innovation, technological dissemination and marketing. Sustainability 6(5):2961–2973

Zhang Y, Nandakumar R, Bartelt-Hunt SL, Snow DD, Hodges L, Li X (2014b) Quantitative proteomic analysis of the Salmonella-lettuce interaction. Microbial Biotechnol 7(6):630–637. https://doi.org/10.1111/1751-7915.12114

Zhang C-x, Tian Y, Cong P-h (2015) Proteome analysis of pathogen-responsive proteins from apple leaves induced by the Alternaria Blotch Alternaria alternata. PLoS One 10(6):e0122233. https://doi.org/10.1371/journal.pone.0122233

Zheng A, Luo J, Meng K, Li J, Zhang S, Li K, Liu G, Cai H, Bryden WL, Yao B (2014) Proteome changes underpin improved meat quality and yield of chickens (Gallus gallus) fed the probiotic Enterococcus faecium. BMC Genomics 15(1):1167. https://doi.org/10.1186/1471-2164-15-1167

Zheng A, Luo J, Meng K, Li J, Bryden WL, Chang W, Zhang S, Wang LXN, Liu G, Yao B (2016) Probiotic (Enterococcus faecium) induced responses of the hepatic proteome improves metabolic efficiency of broiler chickens (Gallus gallus). BMC Genomics 17(1):89. https://doi.org/10.1186/s12864-016-2371-5

Zhu P, Gu H, Jiao Y, Huang D, Chen M (2011) Genomics Proteomics Bioinformatics 9(4):128–137. https://doi.org/10.1016/S1672-0229(11)60016-8

Zhu G, Wu A, Xu X-J, Xiao P, Lu L, Liu J, Cao Y, Chen L, Wu J, Zhao X-M (2015) PPIM: a protein-protein interaction database for maize. Plant Physiol. pp. 01821.02015

Zhu Y, Engström PG, Tellgren-Roth C, Baudo CD, Kennell JC, Sun S, Billmyre RB, Schröder MS, Andersson A, Holm T, Sigurgeirsson B, Wu G, Sankaranarayanan SR, Siddharthan R, Sanyal K, Lundeberg J, Nystedt B, Boekhout T, Dawson TL, Heitman J, Scheynius A, Lehtiö J (2017) Proteogenomics produces comprehensive and highly accurate protein-coding gene annotation in a complete genome assembly of Malassezia sympodialis. Nucleic Acids Res 45(5):2629–2643. https://doi.org/10.1093/nar/gkx006

Chapter 2
Impact of Bioinformatics on Plant Science Research and Crop Improvement

Amrina Shafi, Insha Zahoor, Ehtishamul Haq, and Khalid Majid Fazili

Contents

2.1   Introduction... 29
2.2   Role of Bioinformatics in Crop Improvement................................. 32
2.3   Crop Breeding: Bioinformatics and Preparing for Climate Change............. 35
    2.3.1   Informative Bioinformatics Databases/Tools for Crop Breeders.......... 40
2.4   Application of Bioinformatics in Fruit Breeding............................ 41
2.5   Future Prospects... 42
References... 43

2.1   Introduction

Crops play a significant and diverse role in our economy, environment and feeding the increasing world population. Increased demand for biofuel crops, population explosion and global climate change have become a challenge for current plant biotechnology, and sustainable agricultural production is an urgent issue in this response (Brown and Funk 2008; Ozturk 2010; Hakeem et al. 2012). Climate change will severely influence the world's food supply, and it is predicted to have immense negative effects on both the yield and the quality of crop plants (Kumar 2016),

A. Shafi (✉) · E. Haq
Department of Biotechnology, School of Biological Sciences, University of Kashmir,
Srinagar, Jammu and Kashmir, India
e-mail: haq@kashmiruniversity.ac.in

I. Zahoor
Bioinformatics Centre, University of Kashmir, Srinagar, Jammu and Kashmir, India
e-mail: drinsha@uok.edu.in

K. M. Fazili
Department of Biotechnology, School of Biological Sciences, University of Kashmir,
Srinagar, Jammu and Kashmir, India

Bioinformatics Centre, University of Kashmir, Srinagar, Jammu and Kashmir, India
e-mail: fazili@kashmiruniversity.ac.in

© Springer Nature Switzerland AG 2019   29
K. R. Hakeem et al. (eds.), *Essentials of Bioinformatics, Volume III*,
https://doi.org/10.1007/978-3-030-19318-8_2

unless steps are taken to increase crop resilience. Plant genomics is a potentially powerful defence against this looming threat. Now to solve these issues and increase crop yields, breeding of novel crops and adaptation of current crops to the new environment based on a better molecular understanding of gene function, and on the regulatory mechanisms involved in crop production (Takeda and Matsuoka 2008), have become a primary necessity. Crop yields have increased during the past century and will continue due to enhanced breeding and new biotechnological-engineered strategies. Some of the important gene sequences and their function have been designated, many of which are related to crop yields (production), crop quality, tolerance to biotic and abiotic stresses and development of molecular markers (De Filippis 2012). One vital tool of bioinformatics is "genomics", which is commonly used to identify genotypic and phenotypic changes in plants, and this information helps in improving the overall performance of crop plants (Ahmad et al. 2011).

Modern technologies of bioinformatics have enhanced the study of plant biology to a higher level than before and have assisted in unravelling genetic and molecular networks (Schuster 2007). After a rapid surge in genome sequencing through innovative high-throughput methods, scientists have an opportunity to exploit the structure of the plant genetic material at the molecular level which is known as "plant genomics" (Govindaraj et al. 2015). Some of the latest applications of bioinformatics in plant science research field (Fig. 2.1) are as follows:

- Integrated "omics" strategies clarify the molecular system of the plant which is used to improve the plant productivity. Innovations in omics-based research improve plant-based research.
- Genomics strategy, especially comparative genomics, helps in understanding the genes and their functions and the biological properties of each species.
- Bioinformatics databases are also used in designing new techniques and experiments for increased plant production (Mochida and Shinozaki 2010).
- Advancement in the bioinformatics tools has enabled us in providing information about the genes present in the genome of microorganisms (role in agriculture). These tools have also made it possible to predict the function of different genes and factors affecting these genes, and this information is used by scientists to produce enhanced species of plants which have a drought, herbicide and pesticide resistance in them (Mochida and Shinozaki 2010).
- Nowadays, genomics provides breeders with a new set of tools and techniques that allow the study of the whole genome, and which represents a paradigm shift, by facilitating the direct study of the genotype and its relationship with the phenotype (Tester and Langridge 2010). The present genomics is leading to a new revolution in plant breeding at the beginning of the twenty-first century.

At the most fundamental level, the advances in genomics will greatly accelerate the acquisition of knowledge and that, in turn, will directly affect many aspects of the processes associated with plant improvement. Bioinformatics information and databases have become ready-to-use tools for crop scientists and breeders in gene data mining and linking this knowledge to its biological significance (Mochida and Shinozaki 2010). Knowledge of the function of all plant genes, in

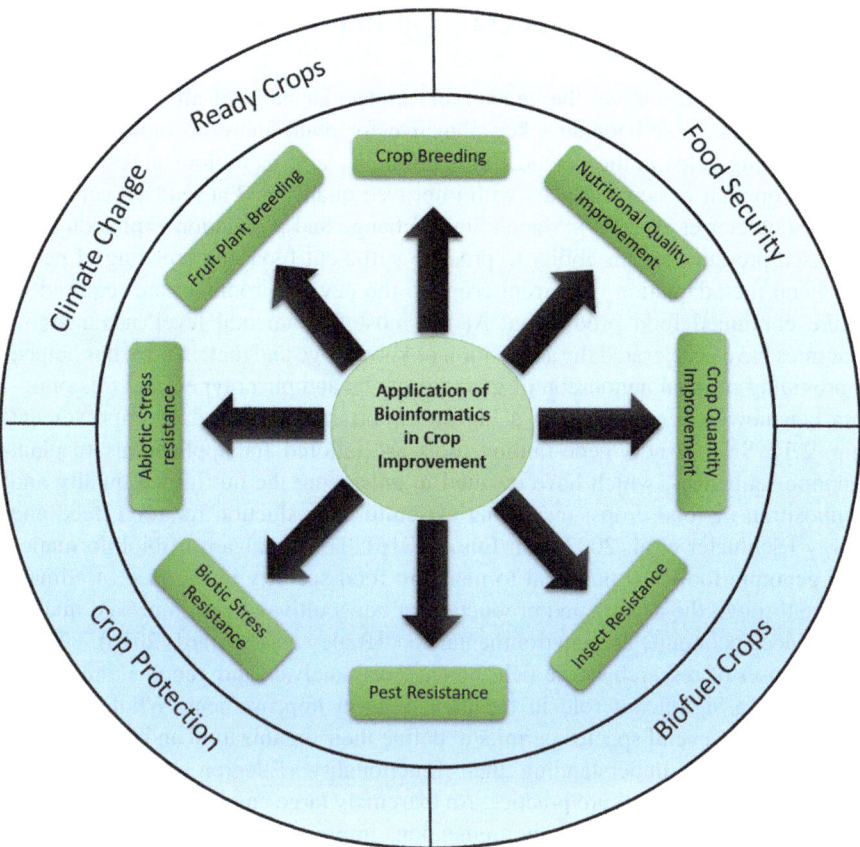

Fig. 2.1 Application of bioinformatics in crop improvement

conjunction with the further development of tools for modifying and interrogating genomes, will lead to the development of a genuine genetic engineering paradigm in which rational changes can be designed and modelled from the first principles. Bioinformatics, when combined with genomics, has the potential to help maintain food security in the face of climate change through the accelerated production of climate-ready crops (Batley and Edwards 2016). Based on these understandings, this chapter focuses on challenges and opportunities, which knowledge and skills in bioinformatics can bring to plant scientists in present plant genomics era as well as future aspects in critical need for effective tools to facilitate the translation of knowledge from new sequencing data to the improvement of plant productivity. This chapter emphasizes on a number of applications of bioinformatics in agriculture in view of crop improvement, breeding programmes, fruit breeding, overviewing the main bioinformatics strategies and challenges, as well as perspectives in this field and various bioinformatics tools/databases important for breeders and plant biotechnologists.

2.2   Role of Bioinformatics in Crop Improvement

To understand and unravel the genetic and molecular basis of all biological processes in plants have become a key objective for plant and crop biologists. This
understanding helps in the practical utilization of plants as biological resources in
the development of new cultivars with improved quality and at reduced economic
costs (Schlueter et al. 2003). Since climate change and population explosion have
increased pressure on our ability to produce sufficient food, the breeding of novel
crops and the adaptation of current crops to the new environment are required to
ensure continued food production. At the most fundamental level, advances in
genomics have accelerated the acquisition of knowledge and that, in turn, has helped
in providing rational annotation of genes, proteins and phenotypes, and this omics
data can now be envisioned as a highly important tool for plant improvement
(Fig. 2.1). Several new gene-finding tools are tailored for applications to plant
genomic sequences, which have resulted in enhancing the nutritional quality and
composition of food crops, increasing agricultural production for food, feed and
energy (Schlueter et al. 2003; Van Emon 2016). This amalgam of bioinformatics
with genomic tools has potential to maintain food security in the face of climate
change through the accelerated production of new cultivars with improved quality
and reduced economic and environmental cost (Batley and Edwards 2016).

The onset of research in the field of sequence analysis and genome annotation
has played a significant role in the area of crop improvement. Whole-genome
sequencing of several species permits to define their organization and provides the
starting point for understanding their functionality (Ellegren 2014), therefore
favouring human agriculture practice. An extremely large amount of genomics data
is available from plants due to the tremendous improvements in the field of omics
(Fig. 2.1), and nowadays function of different genes in the plant and the factors
affecting these genes can be predicted (Morrell et al. 2012). This information has
helped scientists to generate plant species resistant to abiotic and biotic stresses,
herbicides and pesticides. In recent years, a number of latest sequencing technologies, which are adaptations of already existing pyro-sequencing methods (Ansorge
2009), have provided us with new opportunities to be addressed at the entire genome
level in the fields of comparative genomics, meta-genomics and evolutionary
genomics (Varshney et al. 2009). Indeed, the contribution of genomics to agriculture spans the identification and the manipulation of genes linked to specific phenotypic traits (Zhang et al. 2014) as well as genomics breeding by marker-assisted
selection of variants (Organization 2005). Efforts addressed to the achievement of
appropriate knowledge of associated molecular information, such as the one arising
from transcriptome, metabolome and proteome sequencing (Fig. 2.2), are also
essential to better depict the gene content of a genome and its main functionalities.

Current advances in genomics and bioinformatics provide opportunities for
accelerating crop improvement (Fig. 2.2) in the following areas:

- "Gene finding" refers to the prediction of introns and exons in a segment of DNA
  sequence. Bioinformatics has aided in genome sequencing, and it has shown its

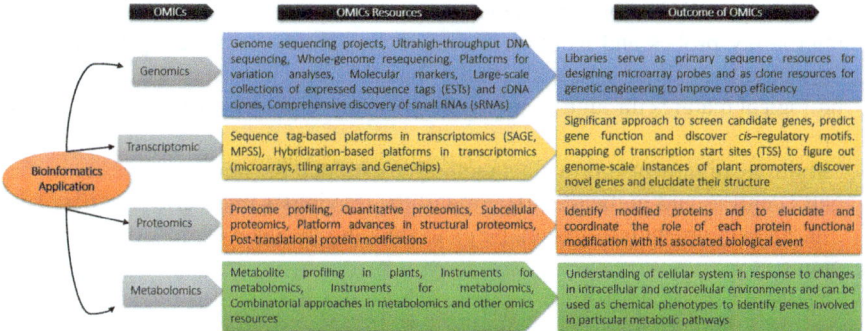

Fig. 2.2 General description of a standard workflow in omics' data analyses

success in locating the genes, in phylogenetic comparison and in the detection of transcription factor binding sites of the genes. Such an approach to identify key genes and understand their function will result in a "quantum leap" in quantitative and qualitative trait improvement in commercially important crops (Morrell et al. 2012).

- "Comparative genetics" (model and non-model plant) with computational tools can reveal an organization of agronomically important genes with respect to each other which can be further used for transferring information from the model crop systems to other food crops. Species-specific nucleotide sequences are now providing information related to phenotypic characters, even when based on genome comparative analyses from the few model plants available (Cogburn et al. 2007; Paterson 2008).

- "Cheminformatics" for designing of agrochemicals is based on an analysis of the components of signal perception and transduction pathways to select targets and to identify potential compounds that can be used as herbicides, pesticides or insecticides, thereby improving plant quality and quantity (Bennetzen et al. 1998).

- "Agricultural genomics" leads to the global understanding of plant and pathogen biology, and its application would be beneficial for agriculture and in providing massive information to improve the crop phenotype. Further, whole-genome sequencing in plants allows chromosome-scale genetic comparisons, thereby identifying conserved genetic areas, which can facilitate identification and documentation of similar genomic sequences in related plant species (Haas et al. 2004; De Bodt et al. 2005).

- "Microarray technology" has been widely adopted in gene expression analysis in crop plants to clarify the function of key genes and uncover the regulation mechanism through dissecting regulatory elements and the interaction of responsible genes. These gene expression studies allow us to understand how plants respond to and interact with the physical environment and management practices. These data may become a crucial tool of future breeding decision management systems (Langridge and Fleury 2011).

- "Full-length cDNA libraries" serve as primary sequence resources for designing microarray probes and as clone resources for genetic engineering to improve crop efficiency (Futamura et al. 2008). These libraries have been used to identify biological features through comparisons of target sequences with those of model organisms.
- "Multiple alignments" provide a method to estimate the number of genes in the gene families and in the identification of the previously undescribed genes. The multiple alignment information helps in studying the gene expression pattern in plants.
- "Mutant analysis" is an effective approach for the investigation of gene function (Stanford et al. 2001). Comprehensive collections of mutant lines are also essential bioresources for radically accelerating forward and reverse genetics.
- "Gene pyramiding" or gene stacking implies multiple desirable genes are assembled from different parent crops to enhance trait and develop elite lines and varieties. It is mainly used in improving existing elite cultivars for a few unsatisfactory traits, for which genes with large positive effects are identified (Malav et al. 2016).
- "Molecular DNA marker" identification and location have contributed significantly to marker-assisted studies and selection (MAS) in plant breeding, and in a wider range of research, including species identification and evolution (Feltus et al. 2004; Varshney et al. 2005).
- "Genetic markers" constructed to cover the complete genome may allow identification of individual genes associated with complex traits by QTL (quantitative trait loci) analysis and the identification of genetic diversity and induced variations (Feltus et al. 2004; Varshney et al. 2005).
- In silico genomics technology has made it easier for researchers (working on plant-pathogen interactions) to identify defence/disease-resistant gene-enzyme with their promoter region and transcription factor which help to enhance the immunity and defence mechanisms (Pandey and Somssich 2009).
- Bioinformatics has also enabled scientists to improve the nutritional quality of the plants by making changes in its genome. Researchers have been successful in inserting genes in the genome of rice to increase vitamin A levels. The genetically modified rice contains more vitamin A (essential to maintain healthy eyes) that has helped in reducing the blindness rate worldwide (Ye et al. 2000).
- Bioinformatics tools are also indispensable to agriculture and horticulture from the climate change perspective. Some varieties of cereal have been modified to be drought/submergence resistant and enhanced to grow in infertile soils.
- "Host-pathogen interactions" help in understanding the disease genetics and pathogenicity factor of a pathogen, which ultimately helped in designing best management options. Metagenomics and transcriptomics approaches are used to understand the genetic architecture of microorganism and pathogens to check how these microbes affect the host plant so that pathogen−/insect-resistant crop is generated and in the identification of host beneficial microbes (Schenk et al. 2012).

- "Insect genomics" helps in the identification of resistance mechanisms and finding the novel target sites (Cory and Hoover 2006). By mapping the genome for *Bacillus thuringiensis* (bacteria that increases soil fertility and protects the plants from pests), scientists were able to incorporate these genes into the plant (e.g. cotton, maize and potato), which made them insect resistant. This resulted in a decrease in insecticide usage, enhancing productivity and nutritional value of crops.
- Improving crops through breeding is a sustainable approach to increase yield and yield stability without intensifying the use of fertilizers and pesticides. Third-generation sequencing technologies are assisting to overcome challenges in plant genome assembly caused by polyploidy and frequent repetitive elements. As a result, high-quality crop reference genomes are increasingly available, benefitting downstream analyses such as variant calling and association mapping that identify breeding targets in the genome (Hu et al. 2018).
- Machine learning also helps identify genomic regions of agronomic value by facilitating functional annotation of genomes and enabling real-time high-throughput phenotyping of agronomic traits in the glasshouse and in the field.
- Crop databases integrate the growing volume of genotype and phenotype data, providing a valuable resource for breeders and an opportunity for data mining approaches to uncover novel trait-associated candidate genes (Hu et al. 2018).

2.3 Crop Breeding: Bioinformatics and Preparing for Climate Change

Plant breeding has been practised for thousands of years, since near the beginning of human civilization (Kingsbury 2009), "plant breeding is the art and science of changing the genetic structure of plants in order to produce desired characteristics" (Sleper and Poehlbman 2006). Bioinformatics has been involved in different aspects of sciences including plant breeding, and a large portion of these tools and techniques are related to the omics category (Barh et al. 2013). From the past few years, plant breeding has been extended through development and deployment of a large number of methods and tools with respect to specific objectives (Al-Khayri et al. 2015). With the use of omics, the consistency and predictability of plant breeding programmes have been improved, reducing the time and the expense of stress-tolerant varieties (Van Emon 2016). The field of genomics and its application to plant breeding are developing very quickly, and this boom in plant breeding has started after genome sequencing of Arabidopsis and rice (Kaul et al. 2000; Matsumoto et al. 2005), followed by many genome sequencing projects of different plant species (Skuse and Du 2008). The combination of conventional breeding techniques with genomic tools and approaches is leading to a new genomics-based plant breeding (Fig. 2.3). A fully assembled and well-annotated genome will allow breeders to discover genes related to agronomic traits, determine their location and function as well as develop genome-wide molecular markers (Hu et al. 2018).

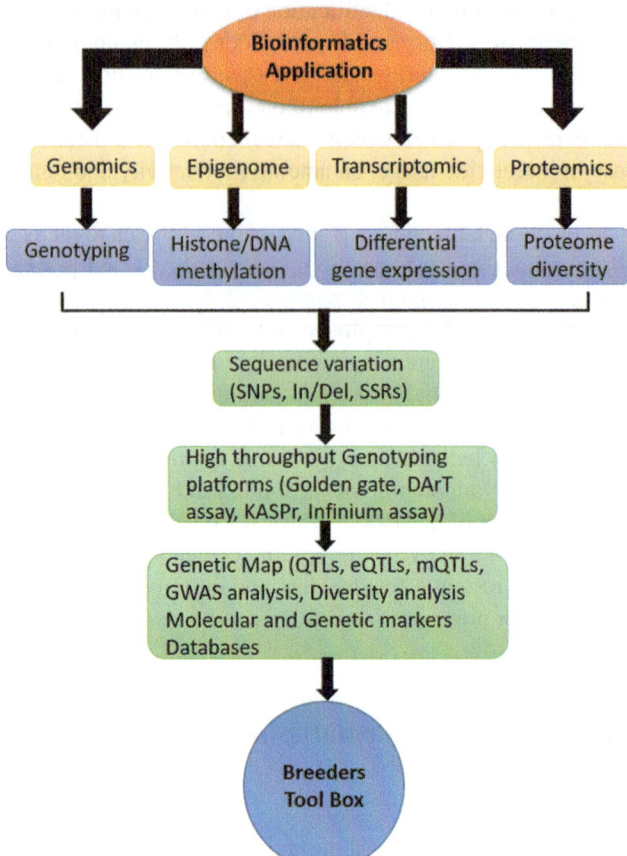

Fig. 2.3 Reaping benefits of omics in crop breeding. Discovery of the genes and the genetic architecture by different omics underlying critical traits provides insights for crop improvement. Identification of genes and quantitative trait loci (QTL) and genome-wide association study (GWAS) enhances rice yield, quality and stress tolerance in a wide range of environments. Genetic maps help to locate genes and provide molecular/genetic markers for selection. Gene discovery provides knowledge of genetic mechanisms and interactions. Databases provided (barcoded) sample tracking and breeding history (pedigrees). Phenotypes (trait measurements) are stored with experimental design and environmental data and can be connected to individual and genotype (marker). All these constitute a toolbox for plant breeders

One of the most substantial transformation of bioinformatics' techniques in plant breeding is that it had replaced the conventional molecular marker technology with high-throughput DNA sequencing technologies and has developed a number of databases (Mochida and Shinozaki 2010) (Table 2.1). These and other technical revolutions provide genome-wide molecular tools for breeders (large collections of markers, high-throughput genotyping strategies, high-density genetic maps, etc.) that can be incorporated into existing breeding methods (Tester and Langridge 2010; Lorenz et al. 2011). With the progress of genome sequencing and large-scale

Table 2.1 GWAS acceleration tools and molecular marker database

S. No.	Databases	Information	Reference
1.	Heap	SNP detection tool for NGS data with special reference to GWAS and detects a larger number of variants taking advantage of the information whether the samples are inbred homozygosity assumption or not	Kobayashi (2015)
2.	GnpIS-Asso	A generic database for managing and exploiting plant genetic association studies. It provides tools that allow plant scientists or breeders to get association values between traits and markers obtained in several association studies. This database is currently used to conduct GWAS on tomato and maize	Steinbach (2015)
3.	BioGPU	A high-performance computing tool for GWAS, BioGPU, effectively controls false positives caused by population structure and unequal relatedness among individuals and improves statistical power when compared to mixed linear model methods	Wang (2015)
4.	BHIT	Bayesian high-order interaction toolkit (BHIT) first builds a Bayesian model on both continuous data and discrete data, which is capable of detecting high-order interactions in SNPs related to case-control or quantitative phenotypes. BHIT effectively detects the high-order interactions associated with phenotypes	Huang (2015)
Databases for molecular and genetic markers			
5.	Plant markers	A genetic marker database that contains predicted molecular markers, such as SNP, SSR and conserved ortholog set (COS) markers, from various plant species	Heesacker et al. (2008)
6.	GrainGenes	A popular site for Triticeae genomics, it provides genetic markers and linkage map data on wheat, barley, rye and oat	Carollo et al. (2005)
7.	Gramene	A database for plant comparative genomics that provides genetic maps of various plant species	Liang et al. (2008)
8.	Triticeae mapped EST database	(TriMEDB) provides information regarding mapped cDNA markers that are related to barley and their wheat homologs	Mochida et al. (2008)
9.	The Panzea	It describes the genetic architecture of complex traits in maize and teosinte. Two common types of markers, SNP and SSR, can be searched	Canaran et al. (2006)
10.	MaizeGDB	A search engine to identify ESTs, AFLPs and RAPD probes and sequence data for maize. The legume information system (LIS) provides access to markers such as SNP, SSR, RFLP and RAPDs for diverse legumes, including peanut, soya bean, alfalfa and common bean	Lawrence et al. (2007)

(continued)

Table 2.1 (continued)

S. No.	Databases	Information	Reference
11.	Sol genomics network database	The data types range from chromosomes and genes to phenotypes and accessions in the Solanaceae and their close relatives. SGN hosts more than 20 genetic and physical maps for tomato, potato, pepper and tobacco with thousands of markers. Genetic marker types in the database include SNP, SSR, AFLP, PCR and RFLP	Bombarely et al. (2011)
12.	SoyBase database	Hosts genomic and genetic data for soya bean. The markers include SNP, SSR, RFLP, RAPD and AFLP	Grant et al. (2010)
13.	MoccaDB	An integrative database for functional, comparative and diversity studies in the Rubiaceae family, which includes coffee. It provides easy access to markers, such as SSR, SNP and RFLP	Plechakova et al. (2009)
14.	The cotton microsatellite database (CMD)	A curated and integrated web-based relational database providing centralized access to publicly available cotton SSRs. CMD contains publication, sequence, primer, mapping and homology data for nine major cotton SSR projects, collectively representing 5484 SSR markers	Blenda et al. (2006)
15.	ICRISAT	A chickpea (*Cicer arietinum* L) root EST database hosted at ICRISAT provides access to over 2800 chickpea ESTs from a library constructed after subtractive suppressive hybridization (SSH)	Jayashree et al. (2005)
16.	SSR primer 2	It provides the real-time discovery of SSRs within submitted DNA sequences, with the concomitant design of PCR primers for SSR amplification. The success of this system has been demonstrated in Brassica and strawberry	Robinson et al. (2004)

EST analysis in various species, these sequence datasets have become quite efficient sequence resources for designing molecular markers covering the entire genomes (Feltus et al. 2004). Recent advances in genomics are producing new plant breeding methodologies, improving and accelerating the breeding process in many ways (e.g. association mapping, marker-assisted selection, "breeding by design", gene pyramiding, genomic selection, etc.) (Lorenz et al. 2011). Some of these molecular and genetic markers, which have played a significant role in improving plant breeding, are as follows:

1. Crop breeders have known the complexity of multiple alleles for decades. However, with the advent of molecular markers, genetic diversity and other forms of genetic structure in breeding populations are possible. For high-throughput genotyping, a number of platforms have been developed that have been applied to genetic map construction, marker-assisted selection and QTL cloning using multiple segregation populations (Hori et al. 2007). Such genotyping systems have also been used in post-genome sequencing projects such as genotyping of genetic resources, accessions to evaluate population structure and association studies to identify genetic loci involved in phenotypic changes of species. Listed in Table 2.1 are the most important web-based sites for DNA markers.

2. Molecular DNA markers have contributed significantly to marker-assisted studies and selection (MAS) in plant breeding and in a wider range of research, including species identification and evolution (Feltus et al. 2004).

3. Genetic markers designed to cover a genome extensively allow not only identification of individual genes associated with complex traits by QTL analysis but also the exploration of genetic diversity with regard to natural variations (Feltus et al. 2004; Varshney et al. 2005).

4. A number of attempts to design polymorphic markers from accumulated sequence datasets have been made for various species, e.g., genome-wide rice (*Oryza sativa*) DNA polymorphism datasets have been constructed based on alignment between *japonica* and *indica* rice genomes (Han and Xue 2003; Shen et al. 2004).

5. The most important database EST (expressed sequence tag) consists of ESTs drawn from the multiple cDNA. Large-scale EST datasets are also important resources for the discovery of sequence polymorphisms, especially for allocating expressed genes onto a genetic map (Heesacker et al. 2008).

6. The Illumina GoldenGate Assay allows the simultaneous analysis of up to 1536 SNPs in 96 samples and has been used to analyse genotypes of segregation populations in order to construct genetic maps allocating SNP (single nucleotide polymorphism) markers in crops (Hyten et al. 2008).

7. Diversity Arrays Technology (DArT) is a high-throughput genotyping system that was developed based on a microarray platform (Wenzl et al. 2007). These DArT markers have been used together with conventional molecular markers to construct denser genetic maps and/or to perform association studies in various crop species.

8. Affymetrix Gene Chip Arrays have been used to discover nucleotide polymorphisms as single-feature polymorphisms based on the differential hybridization of Gene Chip probes in barley and wheat (Rostoks et al. 2005; Bernardo et al. 2009).

9. Transcriptomics subcategories of omics attract a large number of biologists, especially in plant breeding area (Hakeem et al. 2016).

10. The most powerful application of third-generation sequencing for breeding is the assembly of improved contiguous crop genomes (Hu et al. 2018).

Further, as the resolution of genetic maps in the major crops increases, and as the molecular basis for specific traits or physiological responses becomes better elucidated, it will be increasingly possible to associate candidate genes, discovered in model species, with corresponding loci in crop plants (Fig. 2.3). Appropriate relational databases will make it possible to freely associate across genomes with respect to the gene sequence, putative function and genetic map position. Once such tools have been implemented, the distinction between breeding and molecular genetics will fade away. Breeders will routinely use computer models as toolbox to formulate predictive hypotheses to create phenotypes of interest from complex allele combinations (Fig. 2.3) and then construct those combinations by scoring large populations for very large numbers of genetic markers (Walsh 2001).

2.3.1   Informative Bioinformatics Databases/Tools for Crop Breeders

Crop breeding has long relied on cycles of phenotypic selection and crossing, which generate superior genotypes through genetic recombination. When genome sequences are available, all genes and genetic variants contributing to agronomic traits can be identified, and changes made during breeding processes can be assessed at the genotype level (Hu et al. 2018). Availability of ready-to-go genomic data for breeders today plays an increasingly important role in all aspects of crop breeding, such as quantitative trait loci (QTL) mapping and genome-wide association studies (GWAS), where genomic sequencing of crop populations can allow gene-level resolution of agronomic variation. The progress made in genomics-based breeding has even assisted in identification of genetic variation in crop species, which can be applied to produce climate-resilient varieties (Mousavi-Derazmahalleh et al. 2018; Dwivedi et al. 2017).

GWAS (comparative genomic analysis, phylogenomics, evolutionary analysis and genome-wide association study) is presently a favourable tool to explore the allelic variation in a broader scope for extensive phenotypic diversity and higher resolution of QTL mapping. GWAS is an alternative to overcome the disadvantages of existing classical crop breeding methods, e.g. a biparental cross-mapping method for genetic dissections of the agronomically important traits (Myles et al. 2009). GWAS has a powerful application in plant breeding for identifying phenotypic diversity in trait-associated loci, as well as allelic variation in candidate genes addressing quantitative and complex traits (Kumar et al. 2013). GWAS has been successfully applied to study *Arabidopsis thaliana*, where more than 1300 distinct accessions have been genotyped for 250,000 SNP (Kozlov et al. 2015) phenotypes. A few rice genes having large effects in controlling traits are involved in determining yield, morphology and stress tolerance, and nutritional quality was also identified (Famoso et al. 2011). GWAS has been widely used to dissect complex traits in some other major crops, e.g. maize and soya bean (Li et al. 2013; Hwang et al. 2014). Several bioinformatics approaches have been introduced as GWAS acceleration tools (Table 2.1).

Advances in genomics offer the potential to accelerate the genomics-based breeding of crop plants (Fig. 2.3). However, relating genomic data to climate-related agronomic traits for use in breeding remains a huge challenge and one which will require coordination of diverse skills and expertise. Bioinformatics, when combined with genomics, has the potential to help maintain food security in the face of climate change through the accelerated production of climate-ready crops (Batley and Edwards 2016). The vast breeding knowledge gathered over the last several decades will become directly linked to basic plant biology and enhance the ability to elucidate gene function in model organisms (Hospital et al. 2002). The expected dramatic improvements in phenotypes of commercial interest include both the improvement of factors that traditionally limit agronomic performance (input traits) and the alteration of the amount and kinds of materials that crops produce (output traits). Examples include:

- Abiotic stress tolerance
- Biotic stress tolerance
- Improving nutrient use efficiency
- Manipulation of plant architecture and development (size, organ shape, number and position, the timing of development and senescence)
- Metabolite partitioning (redirecting of carbon flow among existing pathways or shunting into new pathways)

Appropriate relational databases will make it possible to freely associate across genomes with respect to the gene sequence, putative function or genetic map position. Once such tools have been implemented, the distinction between breeding and molecular genetics will fade away. Breeders will routinely use computer models to formulate predictive hypotheses to create phenotypes of interest from complex allele combinations and then construct those combinations by scoring large populations for very large numbers of genetic markers (Walsh 2001; Deckers and Hospital 2002).

2.4 Application of Bioinformatics in Fruit Breeding

During the last three decades, the world has witnessed a rapid increase in the knowledge about the plant genome sequences and the physiological and molecular roles of various plant genes, which have revolutionized the molecular genetics and its efficiency in plant breeding programmes. Since bioinformatics has application in every field of science, genome programme can now be envisioned as a highly important tool for fruit breeding. Identifying key genes and understanding their function will result in a "quantum leap" in improving fruit quality and quantity (Meyer and Mewes 2002). The revolution in life sciences brought on by genomics dramatically increases the scale and scope of our experimental enquiry and applications in fruit plant breeding. The scale and high-resolution power of genomics make possible a broad and detailed genetic understanding of plant performance at multiple levels of aggregation (Meyer and Mewes 2002).

The primary goal of fruit plant genomics is to understand the genetic and molecular basis of all biological processes in plants that are relevant to the species. This understanding is fundamental to allow efficient exploitation of fruit plants as biological resources in the development of new cultivars of improved quality and reduced economic and environmental costs (Fig. 2.1). This knowledge is also vital for the development of new diagnostic tools and traits of primary interest like pathogen resistance and abiotic stress, fruit quality and yield. Moreover, gene expression analysis will allow us to understand how fruit plants respond to and interact with the physical environment and management practices. This information, in conjunction with appropriate technology, may provide predictive measures of plant health and fruit quality and become part of future breeding decision management systems. Current genome programmes generate a large amount of data that will require processing, storage and distribution to the international research community. The data include not only sequence information but also information on mutations, markers, maps and functional discoveries.

The key objectives for fruit plant bioinformatics include:

- Integrating phenotypes, genomics and bioinformatics tools and resources in public and private breeding pipelines will address this challenge and help deliver breeding targets
- Providing rational annotation of genes, proteins and phenotypes.
- Elaborating relationships both within the data on individual fruits and between fruits and other organisms.

2.5   Future Prospects

Bioinformatics era and high-throughput sequencing (HTS) are revolutionizing the experimental design in molecular biology, strikingly contributing to increasing scientific knowledge while affecting relevant applications in many different aspects of agriculture. Bioinformatics plays a significant role in the development of the agricultural sector, agro-based industries, agricultural by-product utilization and better management of the environment. With the increase of sequencing projects, bioinformatics continues to make considerable progress in biology by providing scientists with access to the genomic information and plays a big role to analyse the data properly. Recent wealth of plant genomic resources, along with advances in bioinformatics, have enabled plant researchers to achieve a fundamental and systematic understanding of economically important plants and plant processes, critical for advancing crop improvement. The scale and high-resolution power of genomics enable to achieve a broad as well as a detailed genetic understanding of plant performance at multiple levels of aggregation. Advances in genomics are providing breeders with new tools and methodologies that allow a great leap forwards in plant breeding, including the "super domestication" of crops and the genetic dissection and breeding for complex traits. The ability to represent high-resolution physical and genetic maps of crops has been one of the paramount implications of bioinformatics. Plant scientists have an opportunity to use these resources to the full, to ensure that bench work, both in the present and in the future, can be combined with bioinformatics to fully reap the rewards of the genomics revolution. By applying novel technologies and methods in concert, future plant breeding can achieve the crop improvement rate required to ensure food security. Despite these exciting achievements, there remains a critical need for effective tools and methodologies to advance plant biotechnology, to tackle questions that are hardly solved using current approaches and to facilitate the translation of this newly discovered knowledge to improve plant productivity. Overall, due to great impact of plant breeding in order to provide the world food security through improving current staple food crops and also overcome the current harsh environmental situation (as a result of climate change), it is necessary to assess the role and achievements of bioinformatics in breeding science of crop plants.

References

Ahmad P, Ashraf M, Younis M, Hu X, Kumar A, Akram NA, Al-Qurainy F (2011) Role of transgenic plants in agriculture and biopharming. Biotechnol Adv 30:525–540

Al-Khayri JM, Jain SM, Johnson DV (2015) Advances in plant breeding strategies: breeding, biotechnology and molecular tools. Springer International Publishing

Ansorge WJ (2009) Next-generation DNA sequencing techniques. Nat Biotechnol 25:195–203

Barh D, Zambare V, Azevedo V (2013) Omics: applications in biomedical, agricultural, and environmental sciences. CRC Press

Batley J, Edwards D (2016) The application of genomics and bioinformatics to accelerate crop improvement in a changing climate. Curr Opin Plant Biol 30:78–81

Bennetzen JL et al (1998) A plant genome initiative. Plant Cell 10:488–493

Bernardo AN, Bradbury PJ, Ma H, Hu S, Bowden RL, Buckler ES et al (2009) Discovery and mapping of single feature polymorphisms in wheat using Affymetrix arrays. BMC Genomics 10:251

Blenda A, Scheffl er J, Scheffl er B, Palmer M, Lacape JM et al (2006) CMD: a cotton microsatellite database resource for Gossypium genomics. BMC Genomics 7:132

Bombarely A, Menda N, Tecle IY, Buels RM, Strickler S et al (2011) The sol genomics network (solgenomics.Net): growing tomatoes using Perl. Nucleic Acids Res 39:D1149–D1155

Brown M, Funk CC (2008) Climate. Food security under climate change. Science 319:580–581

Canaran P, Stein L, Ware D (2006) LookAlign: an interactive web-based multiple sequence alignment viewer with polymorphism analysis support. Bioinformatics 22:885–886

Carollo V, Matthews DE, Lazo GR, Blake TK, Hummel DD, Lui N et al (2005) Grain genes 2.0. An improved resource for the smallgrains community. Plant Physiol 139:643–651

Cogburn LA, Porter TE, Duclos MJ, Simon J, Burgess SC, Zhu JJ et al (2007) Functional genomics of the chicken-a model organism. Poult Sci 86:2059–2094

Cory JS, Hoover K (2006) Plant-mediated effects in insect-pathogen interactions. Trends Ecol Evol 21:278–286

De Bodt S, Maere S, Van de Peer Y (2005) Genome duplication and the origin of angiosperms. Trends Ecol Evol 20:591–597

De Filippis LF (2012) Breeding for biotic stress tolerance in plants. In: Asharaf M, Ozturk M, Ahmad MSA, Aksoy A (eds) Crop production for agricultural improvement. Springer Science

Deckers J, Hospital F (2002) The use of molecular genetics in the improvement of agricultural populations. Nat Rev Genet 3:22–32

Dwivedi SL, Scheben A, Edwards D, Spillane C, Ortiz R (2017) Assessing and exploiting functional diversity in germplasm pools to enhance abiotic stress adaptation and yield in cereals and food legumes. Front Plant Sci 8:1461

Ellegren H (2014) Genome sequencing and population genomics in nonmodel organisms. Trends Ecol Evol 29(1):51–63

Famoso AN, Zhao K, Clark RT, Tung CW, Wright MH, Bustamante C, Kochian LV, McCouch SR (2011) Genetic architecture of aluminum tolerance in rice (*Oryza sativa*) determined through genome-wide association analysis and QTL mapping. PLoS Genet 7(8):e1002221

Feltus FA, Wan J, Schulze SR, Estill JC, Jiang N, Paterson AH (2004) An SNP resource for rice genetics and breeding based on subspecies *indica* and *japonica* genome alignments. Genome Res 14:1812–1819

Futamura N, Totoki Y, Toyoda A, Igasaki T, Nanjo T, Seki M et al (2008) Characterization of expressed sequence tags from a full-length enriched cDNA library of *Cryptomeria japonica* male strobili. BMC Genomics 9:383

Govindaraj M, Vetriventhan M, Srinivasan M (2015) Importance of genetic diversity assessment in crop plants and its recent advances: an overview of its analytical perspectives. Genet Res Int 20(15):431–487

Grant D, Nelson RT, Cannon SB, Shoemaker RC (2010) SoyBase, the USDA-ARS soybean genetics and genomics database. Nucleic Acids Res 38:D843–D846

Haas BJ, Delcher AL, Wortman JR, Salzberg SL (2004) DAGchainer: a tool for mining segmental genome duplications and synteny. Bioinformatics 20:3643–3646

Hakeem K, Ozturk M, Memon AR (2012) Biotechnology as an aid for crop improvement to overcome food shortage. In: Ashraf M et al (eds) Crop production for agricultural improvement. Springer

Hakeem KR, Tombuloğlu H, Tombuloğlu G (2016) Plant omics: trends and applications. Springer International Publishing, pp 109–136

Han B, Xue Y (2003) Genome-wide intraspecific DNA-sequence variations in rice. Curr Opin Plant Biol 6:134–138

Heesacker A, Kishore VK, Gao W, Tang S, Kolkman JM, Gingle A et al (2008) SSRs and INDELs mined from the sunflower EST database: abundance, polymorphisms, and cross-taxa utility. Theor Appl Genet 117:1021–1029

Hori K, Sato K, Takeda K (2007) Detection of seed dormancy QTL in multiple mapping populations derived from crosses involving novel barley germplasm. Theor Appl Genet 115:869–876

Hospital F, Bouchez A, Lecomete L, Causse M, Charcosset A (2002) Use of markers in plant breeding: lessons from genotype building experiments. 7th WCGALP, Montpellier, pp 22–25

Hu H, Scheben A, Edwards D (2018) Advances in integrating genomics and bioinformatics in the plant breeding pipeline. Agriculture 8(6):75

Huang M (2015) In Biogpu: a high performance computing tool for genome-wide association studies, Plant and Animal Genome XXIII conference, Plant and Animal Genome

Hwang EY, Song Q, Jia G, Specht JE, Hyten DL, Costa J, Cregan PB (2014) A genome-wide association study of seed protein and oil content in soybean. BMC Genomics 15(1):1

Hyten DL, Song Q, Choi IY, Yoon MS, Specht JE, Matukumalli LK et al (2008) High-throughput genotyping with the GoldenGate assay in the complex genome of soybean. Theor Appl Genet 116:945–952

Jayashree B, Buhariwalla HK, Shinde S, Crouch JH (2005) A legume genomics resource: the chickpea root expressed sequence tag database. Electron J Biotechnol 8:128–133

Kaul S, Koo HL, Jenkins J, Rizzo M, Rooney T et al (2000) Analysis of the genome sequence of the flowering plant *Arabidopsis thaliana*. Nature 408(6814):796–815

Kingsbury N (2009) Hybrid: the history and science of plant breeding. University of Chicago Press

Kobayashi M (2015) In heap: a SNPs detection tool for NGS data with special reference to GWAS and genomic prediction, Plant and Animal Genome XXIII conference, Plant and Animal Genome

Kozlov AM, Aberer AJ, Stamatakis A (2015) ExaML version 3: a tool for phylogenomic analyses on supercomputers. Bioinformatics 31(15):2577–2579

Kumar S (2016) Crop breeding: bioinformatics and preparing for climate change. Apple Academic Press

Kumar S, Garrick DJ, Bink MC, Whitworth C, Chagné D, Volz RK (2013) Novel genomic approaches unravel genetic architecture of complex traits in apple. BMC Genomics 14(1):393

Lawrence CJ, Schaeffer ML, Seigfried TE, Campbell DA, Harper LC (2007) MaizeGDB's new data types, resources and activities. Nucleic Acids Res 35:D895–D900

Li H, Peng Z, Yang X, Wang W, Fu J, Wang J, Han Y, Chai Y, Guo T, Yang N (2013) Genome-wide association study dissects the genetic architecture of oil biosynthesis in maize kernels. Nat Genet 45(1):43–50

Liang C, Jaiswal P, Hebbard C, Avraham S, Buckler ES, Casstevens T et al (2008) Gramene: a growing plant comparative genomics resource. Nucleic Acids Res 36:D947–D953

Lorenz AJ, Chao S, Asoro FG, Heffner EL, Hayashi T et al (2011) Genomic selection in plant breeding: knowledge and prospects. Adv Agron 110:77–123

Malav AK, Indu, Chandrawat KS (2016) Gene pyramiding: an overview. Int J Curr Res Biosci Plant Biol 3(7):22–28

Matsumoto T, Wu JZ, Kanamori H, Katayose Y, Fujisawa M et al (2005) The map-based sequence of the rice genome. Nature 436(7052):793–800

Meyer K, Mewes HW (2002) How can we deliver the large plant genomes? Strategies and perspectives. Curr Opin Plant Biol 5:173–177

Mochida K, Shinozaki K (2010) Genomics and bioinformatics resources for crop improvement. Plant Cell Physiol 51:497–523

Mochida K, Saisho D, Yoshida T, Sakurai T, Shinozaki K (2008) TriMEDB: a database to integrate transcribed markers and facilitate genetic studies of the tribe Triticeae. BMC Plant Biol 8:72

Morrell PL, Buckler ES, Ross-Ibarra J (2012) Crop genomics: advances and applications. Nat Rev Genet 13(2):85–96

Mousavi-Derazmahalleh M, Bayer PE, Hane JK, Babu V, Nguyen HT et al (2018) Adapting legume crops to climate change using genomic approaches. Plant Cell Environ 42(1):6–19

Myles S, Peiffer J, Brown PJ, Ersoz ES, Zhang Z, Costich DE, Buckler ES (2009) Association mapping: critical considerations shift from genotyping to experimental design. Plant Cell 21(8):2194–2202

Organization EPS (2005) European plant science: a field of opportunities. J Exp Bot 56(417):1699–1709

Ozturk M (2010) Agricultural residues and their role in bioenergy production. Proceedings-second consultation AgroResidues-Second expert consultation 'The utilization of agricultural residues with special emphasis on utilization of agricultural residues as biofuel', pp 31–43

Pandey SP, Somssich IE (2009) The role of WRKY transcription factors in plant immunity. Plant Physiol 150:1648–1655

Paterson AH (2008) Genomics of sorghum. Int J Plant Genomics 200:362–451

Plechakova O, Tranchant-Dubreuil C, Benedet F, Couderc M, Tinaut A et al (2009) MoccaDB – an integrative database for functional, comparative and diversity studies in the Rubiaceae family. BMC Plant Biol 9:123

Robinson AJ, Love CG, Batley J, Barker G, Edwards D (2004) Simple sequence repeat marker loci discovery using SSR primer. Bioinformatics 20:1475–1476

Rostoks N, Borevitz JO, Hedley PE, Russell J, Mudie S, Morris J et al (2005) Single-feature polymorphism discovery in the barley transcriptome. Genome Biol 6:R54

Schenk PM, Carvalhais LC, Kazan K (2012) Unraveling plant-microbe interactions: can multispecies transcriptomics help? Trends Biotechnol 30:177–184

Schlueter SD, Dong Q, Brendel V (2003) GeneSeqer@PlantGDB: gene structure prediction. Nucleic Acids Res 32:D354–D359

Schuster SC (2007) Next-generation sequencing transforms today's biology. Nature 200(8):16–18

Shen YJ, Jiang H, Jin JP, Zhang ZB, Xi B, He YY et al (2004) Development of genome-wide DNA polymorphism database for map-based cloning of rice genes. Plant Physiol 135:1198–1205

Skuse GR, Du C (2008) Bioinformatics tools for plant genomics. Int J Plant Genomics 2008:910474

Sleper DA, Poehlman JM (2006) Breeding field crops. Blackwell Publishing, Oxford, p 424

Stanford WL, Cohn JB, Cordes SP (2001) Gene-trap mutagenesis: past, present and beyond. Nat Rev Genet 2:756–768

Steinbach D (2015) In GnpIS-Asso: a generic database for managing and exploiting plant genetic association studies results using high throughput genotyping and phenotyping data, Plant and Animal Genome XXIII conference, Plant and Animal Genome

Takeda S, Matsuoka M (2008) Genetic approaches to crop improvement: responding to environmental and population changes. Nat Rev Genet 9:444–457

Tester M, Langridge P (2010) Breeding technologies to increase crop production in a changing world. Science 327:818–822

Van Emon JM (2016) The omics revolution in agricultural research. J Agric Food Chem 64(1):36–44

Varshney RK, Graner A, Sorrells ME (2005) Genomics-assisted breeding for crop improvement. Trends Plant Sci 10:621–630

Varshney RK, Nayak SN, May GD, Jackson SA (2009) Next-generation sequencing technologies and their implications for crop genetics and breeding. Trends Biotechnol 27:522–530

Walsh B (2001) Quantitative genetics in the age of genomics. Theor Popul Biol 59:175–184

Wang J (2015) In A Bayesian model for detection of high-order interactions among genetic variants in genome-wide association studies and its application on soybean oil/protein traits, Plant and Animal Genome XXIII conference, Plant and Animal Genome. p 159–162

Wenzl P, Raman H, Wang J, Zhou M, Huttner E, Kilian A (2007) A DArT platform for quantitative bulked segregant analysis. BMC Genomics 8:196

Ye X, Al-Babili S, Klöti A et al (2000) Engineering the provitamin A (β-carotene) biosynthetic pathway into (carotenoid-free) rice endosperm. Science 287(5451):303–305

Zhang Z, Ober U, Erbe M, Zhang H, Gao N, He J, Li J, Simianer H (2014) Improving the accuracy of whole genome prediction for complex traits using the results of genome wide association studies. PLoS One 9(3):e93017

Chapter 3
Bioinformatics and Plant Stress Management

Amrina Shafi and Insha Zahoor

Contents

3.1   Introduction.. 47
3.2   Plant Genomics-Related Computational Tools and Databases Under Abiotic Stress...... 49
   3.2.1   Genomics Applications in Relation to Abiotic Stress Tolerance................... 51
   3.2.2   Platforms and Resources in the Transcriptome of Plants Under Abiotic
       Stress/Plant Transcriptomics-Related Computational Tools and Databases....... 55
   3.2.3   Platforms and Resources in Proteomic of Plants Under Abiotic Stress/Plant
       Proteomics-Related Computational Tools and Databases......................... 58
   3.2.4   Platforms and Resources in Metabolomics of Plants Under Abiotic
       Stress/Plant Metabolomics-Related Computational Tools and Databases......... 62
   3.2.5   Micro RNAs: Attributes in Plant Abiotic Stress Responses and
       Bioinformatics Approaches on MicroRNA.. 63
3.3   Role of Bioinformatics in Plant Disease Management...................................... 66
3.4   Conclusion and Future Prospects.. 69
References.. 71

3.1   Introduction

In nature, plants are simultaneously exposed to a wide range of stresses (abiotic and biotic), which is a major threat towards the living world more precisely to the plants. This stress leads to various physiological and metabolic changes, which in turn negatively hinder growth, development and productivity of plants (Tardieu and Tuberosa 2010). Based on the climate change study, the occurrence and severity of stresses will surge, resulting in a loss (nearly 70%) of agricultural production (Ghosh and Xu 2014). Thus, an important solution for plant protection and yield

A. Shafi (✉)
Department of Biotechnology, School of Biological Sciences, University of Kashmir, Srinagar, Jammu and Kashmir, India

I. Zahoor
Bioinformatics Centre, University of Kashmir, Srinagar, Jammu and Kashmir, India
e-mail: drinsha@uok.edu.in

© Springer Nature Switzerland AG 2019
K. R. Hakeem et al. (eds.), *Essentials of Bioinformatics, Volume III*,
https://doi.org/10.1007/978-3-030-19318-8_3

increase is by designing plants based on a molecular understanding of gene function and on the regulatory networks involved in stress tolerance, growth and development (Shafi et al. 2014, 2015a, b, 2017). Technological advancement has offered a holistic view on systems organization and functionality; however, the ever-growing extensive data poses great challenges for its efficient analysis and interpretation and finally the integration into different crop improvement schemes (Esposito et al. 2016). Latest, ultrahigh-throughput computational studies are crucial to know about the molecular crosstalks of stress conditions on agricultural crop production. Now the challenge is how to integrate multidimensional biological information in a network and model leading to the development of system biology. Most of the plant system biology strategies rely on four main axes, viz. genomics, proteomics, transcriptomics and metabolomics, which provide us with a better platform to identify and understand the molecular systematics and mechanism under stress conditions (Yuan et al. 2008). Genomics deals with the study of genome; transcriptomics includes structural and functional analyses of coding and non-coding RNA or transcriptome; proteomics deals with protein and post-translational protein modification along with their regulatory pathway and metabolomics, a powerful tool to analyse various metabolites and help in identifying the complex network involved in stress tolerance when analysed in an integrated way. Multifaceted molecular regulatory system and biochemical properties which are specifically involved in stress tolerance and adaptation in plants can be easily deciphered with the help of combined 'omics' study (Chawla et al. 2011). Further, bioinformatics has many practical applications in current plant disease management with respect to the study of host-pathogen interactions, understanding the disease genetics, pathogenicity factor of a pathogen and plant-pathogen biological network, which ultimately help in designing best disease management options (Koltai and Volpin 2003).

Plant amends its 'omics' profiles to cope with the changing environment for their survival, tolerance and growth. The main aim of this 'omic' approach is to find out the molecular interaction and their relationship with the signalling cascade and to process the information which in turn connects specific signals with specific molecular responses (Esposito et al. 2016). The era of genomics, proteomics, metabolomics and phenomics of crop stress biology involves transformation, mining and functional ontology annotation, promoter and SNP analysis, gene expression, pathway enrichment analysis, microRNA prediction, subcellular localization, gene structure analysis, comparative analysis, interactome, protein function analysis, tissues-specific and developmental stage expression analysis and simulation and focused on morpho-molecular differences in stress-exposed and stress-affected crop/model plants. These omics approaches can provide new insights and open new horizons for understanding stresses and responses as well as the improvement of plant responses and resistance to stresses (Duque et al. 2013).

Little is known about the 'omics' characterization of abiotic and biotic stress combinations, but recently, several reports have addressed this issue (Suzuki et al. 2014; Kissoudis et al. 2014). The three main domains that must be addressed to take full advantage of plant systems biology are the development of omics technology, integration of data in a usable format and analysis of data within the domain

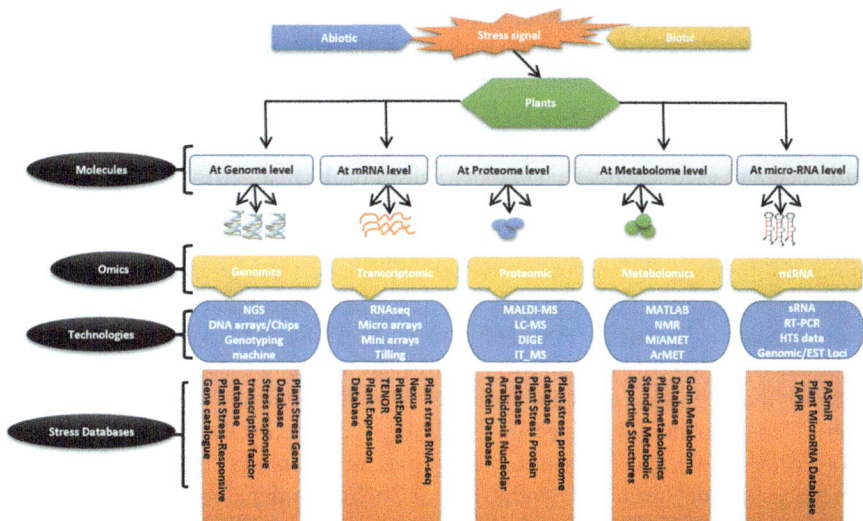

Fig. 3.1 Schematic outline of main 'omics' approaches, their technologies and databases as well as expected outcomes in plant biology and stress research

of bioinformatics. This explicit omics knowledge could subsequently be harnessed by researchers to develop improved crop plants in terms of quality and productivity, showing the enhanced level of abiotic stress tolerance and disease resistance (Singh et al. 2011). In the present chapter, we will introduce the key omics technologies and contemporary innovative technology employed in plant biology and the bioinformatics platforms associated with them (Fig. 3.1). Since the focus of this chapter is the integrated omics approaches in plant stress tolerance, we will describe some of the key concepts, techniques and databases used in bioinformatics, with an emphasis on those relevant to plant stress. It also covers some aspects with regard to the role of application of this endeavour science in today's plant disease management strategies, molecular diagnosis of plant disease in particular to see the application of bioinformatics in detection and diagnosis of plant pathogenic microorganisms.

3.2   Plant Genomics-Related Computational Tools and Databases Under Abiotic Stress

Developments over the past decade, arising predominantly from the human genome project, have led to a new phase of plant genetics known as genomics. 'Genomics' study of all the genes in a given genome includes the identification of gene sequences, intragenic sequences, gene structures and annotations (Duque et al. 2013). This field is the application of the newly available vast amounts of genomic DNA sequence, using a range of novel high-throughput, parallel and other technologies.

The innovation of high-throughput sequencing methods gives scientists the ability to exploit the structure of the genetic material at the molecular level which is known as 'genomics'. Genome sequencing technologies have enabled us to study this part of omics, and it has commenced with the first generation (the 1970s), followed by the next-generation sequencing (NGS) technologies (1990s) as well as the latest third-generation sequencing technologies (El-Metwally et al. 2014b, c). These NGS technologies have got huge impact in plant genome research for the improvement of economically important crops and the understanding of model plant biology. Substantial innovations in platforms for omics-based research and application development provide crucial resources to promote stress-related research in the model and applied plant species (Feuillet et al. 2011). Recent advancement in plant genomics has allowed us to discover and isolate important genes and to analyse functions that regulate yields and tolerance to environmental stress (Govind et al. 2009). Genomics mainly helped in identifying the functional relevance of genes involved in abiotic and biotic stress responses in plants via functional genomic approaches (Ramegowda and Senthil-Kumar 2015). Combinatorial approach using multiple omics platforms and integration of their outcomes is now an effective strategy for clarifying molecular systems integral to improving plant stress tolerance and productivity; this combo approach has helped plant breeders in creating new breeds that can tolerate several biotic and abiotic stresses and, consequently, have increased crop yields as well as pathogen resistance (Shankar et al. 2013; Agarwal et al. 2014). Thus an understanding of plant response towards stresses is enhanced with the application of genomic techniques such as high-throughput analysis of expressed sequence tags (ESTs), large-scale parallel analysis of gene expression, targeted or random mutagenesis and gain-of-function or mutant complementation (Cushman and Bohnert 2000).

Plant genomics study has exploded recently and has become the major boom in plant research due to the rapid increase in plant genomic sequences (Govindaraj et al. 2015). This plant genomic period started from whole-genome sequencing of *Arabidopsis thaliana* (The Arabidopsis Genome Initiative 2000), followed by a draft genome sequence of rice, both *japonica* and *indica* (Yu et al. 2002). Afterwards, the genome sequence of *japonica* rice was completed and published by the International Rice Genome Sequencing Project (International Rice Genome Sequencing Project 2005). Subsequently, the National Science Foundation (NSF) Arabidopsis project (USA) was launched with the stated goal of determining the functions of the 25,000 genes of *Arabidopsis* by 2010. This accumulation of nucleotide sequences of model plants, as well as of applied species such as crops, has provided fundamental information for the design of sequence-based research applications in functional genomics (Somerville and Dangl 2000). Technologies which are included under the canopy of 'genomics' are:

- Automatic DNA sequencing (the machine can read two million base pairs a day)
- Microarrays and DNA chips (tens of thousands of genes can be scanned for activity levels at the same time)
- Automated genotyping machines (assay tens of thousands of DNA diagnostic points a day)

Bioinformatics remains obligatory to projects that seek deciphering of the whole genome of an organism. In fact, soon it will be possible to monitor whole genomes for gene expression on single chips. Once genome sequencing is achieved, one aims to identify and delineate the genomic elements of functional relevance contained within the genome, i.e. 'structural annotation' and assigning biological functions to these elements, referred to as 'functional annotation'.

3.2.1 Genomics Applications in Relation to Abiotic Stress Tolerance

In order to employ applications of genomics field to address the problems of abiotic stress in mandate crops and model plants, approaches like genomic-scale expressed sequence tags (ESTs), genomic sequencing and cDNA microarray analyses have tremendous potentiality in rapidly isolating the candidate genes involved in tolerance mechanisms under stress conditions. Some of the latest techniques used for genomic analysis under stress conditions are as follows:

- Expressed sequence tags (ESTs) are created by partial 'one-pass' sequencing of randomly picked gene transcripts that have been converted into cDNA (Adams et al. 1993). ESTs are often used to be relative collections from stressed and non-stressed plant tissues. A comparison of ESTs of the stressed and non-stressed sample will identify genes that are up-regulated in the stressed tissues and those which are down-regulated or switched off.
- The presence of various key functionalities of full-length cDNA resources in omic space is also essential to establish relevant information resources that provide gateways to these resources as well as to integrate related datasets derived from other omics fields and species (Sakurai et al. 2005).
- cDNA libraries also serve as primary sequence resources for designing microarray probes and as clone resources for genetic engineering to improve crop efficiency (Futamura et al. 2008). Further, candidate genes (induced by stress) which emerge from microarray analyses are ideal for comparative analysis.
- Mini-arrays which are built from collections of ESTs assembled from random cDNA libraries, or from more targeted collections made from cDNAs collected from stressed tissues. Even more targeted will be the special 'stress arrays' made up of all the expressed genes for which there is any evidence of implication.

Omics platforms and their associated databases are also essential for the effective design of approaches making the best use of genomic resources, including resource integration. Various bioinformatics software and tools are being increasingly used to maintain, analyse and retrieve the massive-scale molecular data under stress and non-stressed conditions (Table 3.1). Some of those involved specifically under stress are as follows:

- *Plant Stress Gene Database (PSGD):* It provides information about the genes involved in stress conditions in plants (Prabha et al. 2011). This database includes

Table 3.1 Genomic repositories and stress-related databases

S. No.	Genomics database	Information	References
1.	PlantGDB – resources for Plant Comparative Genomics	Molecular sequence data for all plant species with significant sequencing efforts	Dong et al. (2004)
2.	TAIR (The *Arabidopsis* Information Resource)	Complete genome sequence, gene structure, expression and product information of *Arabidopsis*	Rhee et al. (2003)
3.	Gramene	Online web database resource for plant comparative genomics and pathway analysis based on Ensembl technology	Gupta et al. (2016)
4.	Plant Genome DataBase Japan (PGDBj)	A website that contains information related to genomes of model and crop plants from databases	Nakaya et al. (2017)
5.	PLAZA	Online resource for comparative genomics that integrates plant sequence data and comparative genomic methods	Vandepoele (2017)
6.	Legume Information System (LIS)	Genomic database for the legume family	Dash et al. (2016)
7.	Ensembl Genomes	Analysis and visualization of genomic data and provides access to a variety of data obtained from various sources and analyses, anchored on reference genome sequences	Bolser et al. (2016)
8.	Plant Stress Gene Database	Information about the genes involved in stress conditions in plants	Prabha et al. (2011)
9.	Stress-responsive transcription factor database (STIFDB)	A comprehensive collection of biotic and abiotic stress-responsive genes in *Arabidopsis thaliana* and *Oryza sativa* L. with options to identify probable transcription factor binding sites in their promoters	Shameer et al. (2009)
10.	An Updated Version of Plant Stress-Responsive Transcription Factor database (STIFDB2)	Additional stress signals, stress-responsive transcription factor binding sites and stress-responsive genes in *Arabidopsis* and Rice	Naika et al. (2013)
11.	STIF	Recognition of binding sites of stress-upregulated transcription factors and genes in *Arabidopsis*	Ambika et al. (2008)
12.	Plant Stress-Responsive Gene catalogue (PSRGC)	Database of relationship stress-responsive genes for drought and water with orthologous and paralogous relationships	Wanchana et al. (2018)
13.	PESTD	Database for transcripts with annotated tentative orthologs from crop abiotic stress transcripts	Jayashree et al. (2006)
14.	*Arabidopsis* Stress-Responsive Gene Database	ASRGD database for stress-responsive genes in *Arabidopsis thaliana*	Borkotoky et al. (2013)

(continued)

Table 3.1 (continued)

S. No.	Genomics database	Information	References
15.	Pathogen Receptor Genes Database (PRGDB)	Open and updated space about Pathogen Receptor Genes (PRGs), in which all information available about these genes is stored, curated and discussed	Osuna-Cruz et al. (2018)
16.	Rice SRTFDB	The database which provides information on rice transcription factors drought, salt stress conditions and various developmental stages	Priya and Jain (2013)
17.	QlicRice	A web interface for abiotic stress-responsive QTL and loci interaction channel in rice	Smita et al. (2011)
18.	Drought stress gene database (DroughtDB)	Database for a number of drought stress-associated genes encoding transcription factors that in turn control other various genes involved in diverse physiological and molecular reactions to drought stress	Alter et al. (2015)
19.	MIPS *Oryza sativa* database (MOsDB)	Resource for publicly available sequences of the rice (*Oryza sativa* L.) genome	Wojciech et al. (2003)
20.	MSU rice Genome Annotation Project	Database and Resource is a National Science Foundation project and provides sequence and annotation data for the rice genome	Kawahara et al. (2013)
21.	Plant Genome and Systems Biology (PGSB)	Focuses on the analysis of plant genomes, using bioinformatic techniques, provides a platform for integrative and comparative plant genome research	Spannagl et al. (2016)

259 stress-related genes of 11 species along with all the available information about the individual genes. Stress-related ESTs were also found for *Phaseolus vulgaris*. The database also includes ortholog and paralog of proteins which are coded by stress-related genes.

- *Stress-Responsive Transcription Factor Database (STIFDB V2.0):* It is a comprehensive collection of biotic and abiotic stress-responsive genes in *Arabidopsis thaliana* and *Oryza sativa* L. with options to identify probable transcription factor binding sites in their promoters. In response to biotic stress like bacteria and abiotic stresses like ABA, drought, cold, salinity, dehydration, UV-B, high light, heat, heavy metals, etc., ten specific families of transcription factors in *Arabidopsis thaliana* and six in *Oryza sativa* L. are known to be involved (Shameer et al. 2009).

- *Stress-Responsive Transcription Factor Database (STIFDB2):* Currently it has 38,798 associations of stress signals, stress-responsive genes and transcription factor binding sites predicted using the Stress-responsive Transcription Factor (STIF) algorithm, along with various functional annotation data. As a unique plant stress regulatory genomics data platform, STIFDB2 can be utilized for targeted as well as high-throughput experimental and computational studies to unravel principles of the stress regulome in dicots and gramineae (Naika et al. 2013).

- *STIF (Hidden Markov Model-Based Search Algorithm):* It is used for the recognition of binding sites of stress-upregulated transcription factors and genes in *Arabidopsis* (Ambika et al. 2008).
- *PESTD:* A comparative genomics study on plant responses to abiotic stresses and is a dataset of orthologous sequences. A large amount of sequence information, including those derived from stress cDNA libraries, are used for the identification of stress-related genes and orthologs associated with the stress response. Availability of annotated plant abiotic stress ortholog sets will be a valuable resource for researchers studying the biology of environmental stresses in plant systems, molecular evolution and genomics (Jayashree et al. 2006).
- *Arabidopsis Stress-Responsive Gene Database (ASRGD):* It is a powerful mean for manipulation, comparison, search and retrieval of records describing the nature of various stress-responsive genes in *Arabidopsis thaliana*. About 44 types of different stress factors are related to *Arabidopsis thaliana*, and the database contains 636 gene entries related to stress response with their related information like gene ID, nucleotide and protein sequences and cross-response. The database is based exclusively on published stress-responsive and stress-tolerant genes associated with plants (Borkotoky et al. 2013).
- *The Arabidopsis Information Resource (TAIR):* It contains genetic and molecular biology data for the *Arabidopsis thaliana*, which is more widespread to different aspects apart from the stress response, which makes it difficult to look for only stress-related genes (Swarbreck et al. 2008).
- *Pathogen Receptor Genes Database (PRGDB):* It allows easy access not only to the plant science research community but also to breeders who want to improve plant disease resistance. It offers 153 reference resistance genes and 177,072 annotated candidate pathogen receptor genes (PRGs). Plant diseases display useful information linked to genes and genomes to connect complementary data and better address specific needs. Through a revised and enlarged collection of data, the development of new tools and a renewed portal, PRGdb 3.0 engages the plant science community in developing a consensus plan to improve knowledge and strategies to fight diseases that afflict main crops and other plants (Osuna-Cruz et al. 2018).
- *Rice SRTFDB:* It provides comprehensive expression information on rice transcription factors (TFs) during drought and salinity stress conditions and various stages of development. It will be useful to identify the target TF(s) involved in stress response at a particular stage of development. It also provides curated information for *cis*-regulatory elements present in their promoters, which will be important to study the binding proteins. This database aims to accelerate functional genomics research of rice TFs and understand the regulatory mechanisms underlying abiotic stress responses (Priya and Jain 2013).
- *QlicRice:* This database is designed to host publicly accessible, abiotic stress-responsive quantitative trait loci (QTLs) in rice (*Oryza sativa*) and their corresponding sequenced gene loci. It provides a platform for the data mining of abiotic stress-responsive QTLs, as well as browsing and annotating associated traits, their location on a sequenced genome, mapped expressed sequence tags

(ESTs) and tissue- and growth stage-specific expressions on the whole genome. An appropriate and spontaneous user interface has been designed to retrieve associations to agronomically important QTLs on abiotic stress response in rice (Smita et al. 2011).

- *Drought Stress Gene Database (DroughtDB):* It is a manually curated compilation of molecularly characterized genes that are involved in drought stress response. It includes information about the originally identified gene, its physiological and/or molecular function and mutant phenotypes and provides detailed information about computed orthologous genes in nine model and crop plant species. Thus, DroughtDB is a valuable resource and information tool for researchers working on drought stress and will facilitate the identification, analysis and characterization of genes involved in drought stress tolerance in agriculturally important crop plants (Alter et al. 2015).

3.2.2 Platforms and Resources in the Transcriptome of Plants Under Abiotic Stress/Plant Transcriptomics-Related Computational Tools and Databases

Transcriptome (RNA sequencing or expression profile of an organism) is highly dynamic and involves capturing of the RNA expression profile in spatial and temporal plant organs, tissues and cells within particular conditions (Duque et al. 2013; El-Metwally et al. 2014a). In response to various abiotic stresses, the plant constantly adjusts their transcriptome profile. Thus, transcriptomics study assists in finding genes that are associated with alterations in the plant phenotype under different abiotic or biotic stress conditions (Kawahara et al. 2013). This comprehensive and high-throughput RNAseq analysis finds its applications in plant stress response and tolerance such as searching for abiotic stress candidate genes, predicting tentative gene functions, discovering *cis*-regulatory motifs and providing a better understanding of the plant-pathogen relationship (De Cremer et al. 2013; Agarwal et al. 2014). The recent boom in the availability of online resources, databases and archives of transcriptome data allows for performing novel genome-wide analysis of plant stress responses and tolerance (Duque et al. 2013). Several studies on the transcriptome of different organs and developmental stages of plants under different environmental conditions were observed (Narsai et al. 2010; Zhou et al. 2008). Narsai et al. (2010) identified an exclusively new set of reference genes in rice that are of immense significance, and analysis of their promoter sequence shows the prevalence of some stress regulatory *cis*-element (Zhou et al. 2008).

Different techniques exist to analyse transcriptomic changes in a system under different stress conditions and these are as follows:

- RNA/gene expression profiling is mostly accomplished using microarray, RNA sequencing (RNAseq) through next-generation sequencing (NGS), serial analysis of gene expression (SAGE) and digital gene expression profiling (Kawahara et al. 2013; Duque et al. 2013; De Cremer et al. 2013).

- Hybridization-based method, such as that used in microarrays and GeneChips, has been well established for acquiring large-scale gene expression profiles for various species (De Cremer et al. 2013).
- Next-generation DNA sequencing application, deep sequencing of short fragments of expressed RNAs, including sRNAs, is quickly becoming an effective tool for use with genome-sequenced species (Harbers and Carninci 2005).
- Quantitative PCR analyses up to a few genes at a time, while microarray analysis allows the simultaneous measurement of transcript abundance for thousands of genes (Joshi et al. 2012).
- Tiling arrays cover the genome at regular intervals to measure transcription without bias towards known or predicted gene structures, the discovery of polymorphisms, analysis of alternative splicing and identification of transcription factor binding sites (Coman et al. 2013). Transcriptome analysis in *Arabidopsis* under abiotic stress conditions using a whole-genome tiling array resulted in the discovery of antisense transcripts induced by abiotic stresses (Matsui et al. 2008).

In the post-genomic era, RNA-Seq provides a global transcriptome profile, which could cover lncRNAs, coding genes and their alternatively spliced isoforms in stress response, and aids plant biologists to expand new insights into molecular mechanisms and responses to biotic and abiotic stress events. Several data portals contain a vast amount of plant RNA-Seq data, such as the National Center for Biotechnology Information (NCBI) Gene Expression Omnibus (GEO) and the Sequence Read Archive (SRA). However, these data portals mainly serve as raw biological data archives. Large-scale stress-specific RNA-Seq database that can provide comprehensively visualized transcriptome expression profiles and statistical analysis for differential expression has been listed in Table 3.2. Some of these databases are as follows:

- *Plant Stress RNA-Seq Nexus (PSRN):* It is a comprehensive database which includes 12 plant species, 26 plant stress RNA-Seq datasets and 937 samples. PSRN is an open resource for intuitive data exploration, providing expression profiles of coding-transcript/lncRNA and identifying which transcripts are differentially expressed between different stress-specific subsets, in order to support researchers generating new biological insights and hypotheses in molecular breeding or evolution. PSRN was developed with the goal of collecting, processing, analysing and visualizing publicly available plant RNA-Seq data (Li et al. 2018).
- *PlantExpress:* It is a web database as a platform for gene expression network (GEN) analysis with the public microarray data of rice and *Arabidopsis*. PlantExpress has two functional modes: single-species mode is specialized for GEN analysis within one of the species, while the cross-species mode is optimized for comparative GEN analysis between the species. It stores data obtained from three microarrays, namely, the Affymetrix Rice Genome Array, the Agilent

Table 3.2 Transcriptomic repositories and stress-related databases

S. No.	Transcriptomic database	Information	References
1.	Plant Stress RNA-Seq Nexus (PSRN)	Stress-specific transcriptome database in plant cells	Li et al. (2018)
2.	TENOR (Transcriptome Encyclopedia Of Rice)	Database for comprehensive mRNA-Seq experiments in rice	Kawahara et al. (2016)
3.	PlantExpress	A database integrating OryzaExpress and ArthaExpress for single-species and cross-species gene expression network analyses with microarray-based transcriptome data	Kudo et al. (2017)
4.	RiceArrayNet	A database for correlating gene expression from transcriptome profiling, and its application to the analysis of co-expressed genes in rice	Lee et al. (2009)
5.	PLEXdb (Plant Expression Database)	Unified gene expression resource for plants and plant pathogens	Dash et al. (2012)
6.	EGENES	Transcriptome-based plant database of genes with metabolic pathway information and expressed sequence tag indices in KEGG, provides gene indices for each genome	Masoudi-Nejad et al. (2007)
7.	MOROKOSHI	Transcriptome database in *Sorghum bicolor*	Makita et al. (2015)
8.	AgBase	Genome-wide structural and functional annotation and modelling of microarray and other functional genomics data in agricultural species	McCarthy et al. (2006)
9.	Tiling arrays	Measure transcription without bias towards known or predicted gene structures, the discovery of polymorphisms, analysis of alternative splicing and identification of transcription factor binding sites	Coman et al. (2013)
10.	Chickpea Transcriptome Database (CTDB)	Comprehensive information about the chickpea transcriptome. The database contains various information and tools for transcriptome sequence, functional annotation, conserved domain(s), transcription factor families, molecular markers (microsatellites and single nucleotide polymorphisms)	Verma et al. (2015)
11.	TodoFirGene	Omics information of gymnosperms and connect researchers from forest sciences with those in comparative bioinformatics and evolutionary sciences	Ueno et al. (2018)
12.	ROSAcyc	Resource pathway database that allows access to the putative genes and enzymatic pathways, provides useful information on *Rosa*-expressed genes, with thorough annotation and an overview of expression patterns for transcripts with good accuracy	Dubois et al. (2012)

(continued)

Table 3.2 (continued)

S. No.	Transcriptomic database	Information	References
13.	KEGG PATHWAY Database (Univ. of Kyoto)	Database resource for understanding high-level functions and utilities of the biological system, such as the cell, the organism and the ecosystem, from molecular-level information, especially large-scale molecular datasets generated by genome sequencing and other high-throughput experimental technologies	Kanehisa and Goto (2000)

Rice Gene Expression 4x44K Microarray and the Affymetrix *Arabidopsis* ATH1 Genome Array, with respective totals of 2,678, 1,206 and 10,940 samples. PlantExpress will facilitate understanding of the biological functions of plant genes (Kudo et al. 2017).

- *RiceArrayNet (RAN):* It provides information on co-expression between genes in terms of correlation coefficients (*r* values). A correlation pattern between Os01g0968800, a drought-responsive element-binding transcription factor; Os02g0790500, a trehalose-6-phosphate synthase; and Os06g0219500, a small heat shock factor, reflecting the fact that genes responding to the same biological stresses is regulated together (Lee et al. 2009).

- *Transcriptome Encyclopedia of Rice (TENOR):* It is a database that encompasses large-scale mRNA sequencing (mRNA-Seq) data obtained from rice under a wide variety of stress conditions. Since the elucidation of the ability of plants to adapt to various growing conditions is a key issue in plant sciences, it is of great interest to understand the regulatory networks of genes responsible for environmental changes. All the resources (novel genes identified from mRNA-Seq data, expression profiles, co-expressed genes and *cis*-regulatory elements) are available in TENOR (Kawahara et al. 2016).

3.2.3 Platforms and Resources in Proteomic of Plants Under Abiotic Stress/Plant Proteomics-Related Computational Tools and Databases

'Proteome' referred to the total expressed protein under certain circumstances in a given organism, organ, cell, tissue or microorganism population, and it comprises all the techniques used in profiling the expressed proteins in a specific context (Tyers and Mann 2003). Similar to the transcriptome, it is an informative approach used to reveal invaluable information when studying plant stress response and tolerance, either in a whole genome or sample scale (Nakagami et al. 2012). It is used for profiling all the expressed proteins under multiple stress conditions and cross-comparing these different sets to identify the proteins which are specifically involved in stress tolerance (Yan et al. 2014). This is an evolving technology for the qualitative large-scale identification and quantification of all protein types in a cell or tissue, analysis of post-translational modifications and association with other proteins, and characterization of protein activities and structures (Jorrín-Novo et al. 2009).

Proteomics is associated with two types of studies: proteome characterization (identification of all the proteins expressed) and differential proteomics (comparative proteome analysis of control and stressed plants). The proteomic approach has been largely adopted to explore the protein profiles in plants in response to abiotic stress that might lead to the development of new strategies for improving stress tolerance (Helmy et al. 2011). Several types of proteomes can be measured, but whole proteome and the phosphoproteome are the most common proteomes quantified in plant stress tolerance (Helmy et al. 2011, 2012a, b). The main focus of quantitative proteomics is to identify the proteins that are differentially expressed under certain stress response condition (Liu et al. 2015), while phosphoproteomics is closely associated with the identification of proteins activated and functioning in response to particular stress (Zhang et al. 2014). Both whole proteomics and phosphoproteomics can be combined in one comprehensive study to provide a better understanding of the stress (Hopff et al. 2013). The main goal of functional proteomics is the high-throughput identification of all proteins that appeared in cells and/or tissues, but recent rapid technical advances in proteomics have enabled us to progress to the second generation of functional proteomics, including quantitative proteomics, subcellular proteomics and various modifications and protein-protein interactions (Jorrín-Novo et al. 2009).

Two main techniques that are mostly used for quantitative and/or qualitative profiling are protein electrophoresis and protein identification with mass spectrometry. The technology of choice for proteomics is mass spectrometry (MS) including several approaches such as liquid chromatography-mass spectrometry (LC-MS/MS), ion trap-mass spectrometry (IT-MS) and matrix-assisted laser desorption/ionization-mass spectrometry (MALDI-MS) (Helmy et al. 2011, 2012a). These technologies are basically used in measuring the mass and charge of small protein fragments (or 'peptides') that result from protein enzymatic digestion (Helmy et al. 2011; Nakagami et al. 2012). Furthermore, several proteomics labs use protein electrophoresis technologies such as two-dimensional electrophoresis and difference gel electrophoresis (DIGE) in plant proteomics (Duque et al. 2013).

As genome sequencing projects for several organisms have been completed, proteome analysis, which is the detailed investigation of the functions, functional networks and 3D structures of proteins, has gained accumulative consideration. Large-scale proteome datasets available serve as an imperative resource for a better understanding of protein functions in cellular systems, which are controlled by the dynamic properties of proteins (Table 3.3). These properties reflect cell and organ states in terms of growth, development and response to environmental changes. Functional and experimental validation of proteins associated with biotic and abiotic stresses has been employed as the sole criterion for inclusion in the database (Singh et al. 2015). Due to the challenges faced in text/data mining, there is a large gap between the data available to researchers and the hundreds of published plant stress proteomics articles. There are a large number of stress-related databases for proteins (Table 3.3):

Table 3.3 Proteomic databases and resources

S. No.	Proteomic database	Information	References
1.	Protein-Protein Interaction Inhibition Database (2PI2db)	A database containing the structures of protein-protein and protein-modulator complexes that have been characterized by X-ray crystallography or NMR	Basse et al. (2016)
2.	PPDB is a Plant Proteome DataBase	National science foundation-funded project to determine the biological function of each protein in plants. It includes data for two plants that are widely studied in molecular biology, *Arabidopsis thaliana* and maize (*Zea mays*)	Sun et al. (2009)
3.	Plant stress proteome database (PlantPReS)	Enables researchers to perform multiple analyses on the database; the results of each query indicate a series of proteins for which a set of selected criteria are met	Mousavi et al. (2016)
4.	*Arabidopsis* Nucleolar Protein Database (AtNoPDB)	Information on the plant proteins identified to date with a comparison to human and yeast proteins and images of cellular localizations for over a third of the proteins	Brown et al. (2005)
5.	Plant Protein Phosphorylation Database (P³DB)	Integrated database for plant protein phosphorylation can help identify functionally conserved phosphorylation sites in plants using a multi-system approach	Gao et al. (2009)
6.	PhosPhAt (Plant Protein Phosphorylation DataBase)	A database that specifically maintains experimental phosphorylation site data for *Arabidopsis*	Heazlewood et al. (2008) and Durek et al. (2010)
7.	PSPDB (Plant Stress Protein Database)	Current plant stress databases report plant genes without protein annotations specific to these stresses	Singh et al. (2015)
8.	MANET database Molecular Ancestry Network	The database is a bioinformatics database that maps evolutionary relationships of protein architectures directly onto biological networks	Kim et al. (2006)
9.	PhytAMP	Database of plant natural antimicrobial peptides	Hammami et al. (2009)
10.	PRINTS	Compendium of protein fingerprints	Attwood et al. (1994)
11.	PROSITE	Database of protein families and domains	Hulo et al. (2006)
12.	Swiss-Prot	Protein knowledgebase	Bairoch and Apweiler (2000)
13.	Protein DataBank in Europe (PDBe)	European part of the wwPDB for the collection, organization and dissemination of data on biological macromolecular structures	Mir et al. (2018)

(continued)

Table 3.3 (continued)

S. No.	Proteomic database	Information	References
14.	ProteinDatabank in Japan (PDBj)	Maintains a centralized PDB archive of macromolecular structures and provides integrated tools	Kinjo et al. (2017)
15.	Research Collaboratory for Structural Bioinformatics (RCSB)	Protein Data Bank archive information about the 3D shapes of proteins, nucleic acids and complex assemblies that helps students and researchers understand all aspects of biomedicine and agriculture	Rose et al. (2017)
16.	Structural Classification of Proteins (SCOP)	The database is a largely manual classification of protein structural domains based on similarities of their structures and amino acid sequences	Murzin et al. (1995)
17.	InterPro	Classifies proteins into families and predicts the presence of domains and sites	Hunter et al. (2009)
18.	Pfam	Protein families database of alignments and HMMs	Finn et al. (2013)

- *Plant Stress Proteome Database (*PlantPReS*; www.proteome.ir):* It is an open online proteomic database, which currently comprises >35,086 entries from 577 manually curated articles and contains >10,600 unique stress-responsive proteins (Mousavi et al. 2016).
- *Plant Stress Protein Database (PSPDB):* It is one of the largest repositories and a web-accessible resource that covers 2064 manually curated plant stress proteins from a wide array of 134 plant species with 30 different types of biotic and abiotic stresses. Functional and experimental validation of proteins associated with biotic and abiotic stresses has been employed as the sole criterion for inclusion in the database (Singh et al. 2015).

'Proteogenomics' is another comprehensive combo approach of large-scale proteomic data with genomic and/or transcriptomics data to elucidate various innovative regulatory mechanisms (Helmy et al. 2012a). The proteomics data generated by means of MS-based proteomics (high throughput and accuracy) provides a rich source of translation-level information about the expressed proteins that can be used as a source of large-scale experimental evidence for several predictions (Helmy et al. 2012a, b). In a proteogenomics study, the naturally expressed proteins are identified using MS-based proteogenomics followed by mapping them back to the genomic or transcriptomic data (Helmy et al. 2012a). This field has facilitated in elevating our understanding of the biology of plants in general as well as plant stress research in particular. For instance, a large-scale proteogenomics study of *Arabidopsis thaliana* identified 57 new genes and corrected the annotations of hundreds of its genes using intensive sampling from the *Arabidopsis* organs under several conditions (Baerenfaller et al. 2008). Another study reported corrections and new identifications in about 13% of the annotated genes in *Arabidopsis* (Castellana et al. 2008). It also gives information on the investigation of the host-pathogen relationship (Delmotte et al. 2009), identifying novel effectors in fungal diseases (Cooke et al. 2014), as well as shedding light on the mechanisms of environmental adaptation.

3.2.4   Platforms and Resources in Metabolomics of Plants Under Abiotic Stress/Plant Metabolomics-Related Computational Tools and Databases

The metabolome is the complete pool of metabolites in a cell at any given time and metabolomics refers to techniques and methods used to study the metabolome (Duque et al. 2013). Plants are able to synthesize a diverse group of chemical and biological compounds with different biological activity that is crucial for regulating the response to different types of biotic and abiotic stress (Bino et al. 2004). Therefore, identifying the metabolites produced by the plant under each stress condition by metabolomics plays a significant role to gather information not only about the phenotype but changes in it induced by stress, thereby bridging the gap between phenotype and genotype (Badjakov et al. 2012). Metabolomics may prove to be particularly important in plants due to its ability to elucidate plant cellular systems and permits engineering molecular breeding to improve the growth and productivity of plants in stress tolerance (Fernie and Schauer 2009). Metabolomic approaches allow us to conduct parallel assessments of multiple metabolites, and it is notable that the plant metabolome represents an enormous chemical diversity due to the complex set of metabolites produced in each plant species (Bino et al. 2004). A strong connection between stress metabolites and a particular protein indicates the role of this gene in the stress response process (Urano et al. 2010; Duque et al. 2013; Jogaiah et al. 2013). Metabolic profiling of plants involves a combo of several analytical, separation techniques and with other omics analysis (e.g. transcriptomics or proteomics) to investigate the correlation between metabolite levels and the expression level of genes/proteins (Jogaiah et al. 2013). Thus, metabolomics provides a better understanding of the stress response and tolerance process in model plants such as *Arabidopsis* (Cook et al. 2004) as well as in crops like a common bean (*Phaseolus vulgaris*) (Broughton et al. 2003), and other food crops (Duque et al. 2013).

This is one of the most rapidly developing technologies, and many notable technological advances have recently been made in instrumentation related to metabolomics; some of them are as follows:

- Major approaches that are used in plant metabolomics research include metabolic fingerprinting which involves separation of metabolites based on the physical and chemical properties using various analytical tools and technologies (Jogaiah et al. 2013).
- Metabolite profiling which includes the study of the alterations in metabolite pool that are induced by stress and finally target analyses.

- Capillary electrophoresis-liquid chromatography-mass spectrometry (CE-MS) is considered the most advanced metabolomics technology (Soga et al. 2002).
- Analytical instruments and separation technologies are employed in metabolomics such as gas chromatography (GC), mass spectrometry (MS) and nuclear magnetic resonance (NMR) (Duque et al. 2013).
- Metabolomics experiment (MIAMET) gives reporting requirements with the aim of standardizing experiment descriptions, particularly within publications (Ernst et al. 2014).
- Standard Metabolic Reporting Structures (SMRS) working group has developed standards for describing the biological sample origin, analytical technologies and methods used in a metabolite profiling experiment (Chen et al. 2015).
- ArMet (architecture for metabolomics) proposal gives a description of plant metabolomics experiments and their results along with a database schema (Castillo-Peinado and de Castro 2016).
- Metabolic flux analysis measures the steady-state flow between metabolites. FluxAnalyzer is a package for MATLAB that integrates pathway and flux analysis for metabolic networks (Rocha et al. 2008).

A number of studies of metabolic profiling in plant species have been performed that have resulted in the publication of related databases (Table 3.4). For instance, metabolic pathways that act in response to environmental stresses in plants were investigated by metabolome analysis using various types of MS coupled with microarray analysis of overexpressors of genes encoding two TFs, DREB1A/CBF3 and DREB2A (Maruyama et al. 2009). Metabolomic profiling was also used to investigate chemical phenotypic changes between wild-type *Arabidopsis* and a knockout mutant of the *NCED3* gene under dehydration stress conditions (Urano et al. 2010). These databases are vast information resources and repositories of large-scale datasets and also serve as tools for further integration of metabolic profiles containing comprehensive data acquired from other omics research (Akiyama et al. 2008). One of the huge databases for metabolites is PlantMetabolomics.org (PM), which is a web portal and database for exploring, visualizing and downloading plant metabolomics data. Widespread public access to well-annotated metabolomics datasets (Table 3.4) is essential for establishing metabolomics as a functional genomics tool. PM can be used as a platform for deriving hypotheses by enabling metabolomic comparisons between genetically unique *Arabidopsis* (*Arabidopsis thaliana*) populations subjected to different environmental conditions (Bais et al. 2015).

3.2.5 Micro RNAs: Attributes in Plant Abiotic Stress Responses and Bioinformatics Approaches on MicroRNA

MicroRNA (miR) represents a major subfamily of endogenously transcribed sequences (21–24 bp) and has been acknowledged as a major regulatory class that inhibits gene expression in a sequence-dependent manner (Eldem et al. 2013). miRs

Table 3.4 Metabolomic databases and resources

S. No.	Metabolomics database	Information	References
1.	Metabolome Tomato Database (MoTo DB)	LC-MS-based metabolome database	Moco et al. (2006)
2.	KOMICS (Kazusa-omics) database	Annotations of metabolite peaks detected by LC-FT-ICR-MS and containing a representative metabolome dataset for the tomato cultivar (Micro-tom)	Iijima et al. (2008)
3.	Golm Metabolome Database (GMD)	Provides public access to custom mass spectra libraries and metabolite profiling experiments as well as to additional information and related tools	Kopka et al. (2005)
4.	MS/MS spectral tag (MS2T) libraries at the Platform for Riken Metabolomics (PRIMe) website	Provides access to libraries of phytochemical LC-MS2 spectra obtained from various plant species by using an automatic MS2 acquisition function of LC-ESI-Q-TOF/MS	Matsuda et al. (2009)
5.	Armec Repository Project	Metabolome data on the potato and serves as a data repository for metabolite peaks detected by ESI-MS	Gomez-Casati et al. (2013)
6.	Plant metabolomics	The NSF-funded multi-institutional project aimed at the development of the *Arabidopsis* metabolomics database	Bais et al. (2015)
7.	Minimum information about a metabolomics experiment (MIAMET)	Reporting requirements with the aim of standardizing experiment descriptions, particularly within publications	Ernst et al. (2014)
8.	Standard Metabolic Reporting Structures (SMRS)	Standards for describing the biological sample origin, analytical technologies and methods used in a metabolite profiling experiment	Chen et al. (2015)
9.	ArMet (architecture for metabolomics)	The proposal gives a description of plant metabolomics experiments and their results along with a database schema	Castillo-Peinado and de Castro (2016)
10.	MATLAB (Metabolic flux analysis) FluxAnalyzer	Measures the steady-state flow between metabolites. It is a package that integrates pathway and flux analysis for metabolic networks	Rocha et al. (2008)
11.	BiGG Models	Knowledge base of genome-scale metabolic network reconstructions, integrating more than 70 published genome-scale metabolic networks into a single database with a set of standardized identifiers called BiGG IDs	King et al. (2016)
12.	BioCyc Database Collection	Microbial genome web portal that combines thousands of genomes with additional information inferred by computer programmes, imported from other databases, and curated from the biomedical literature by biologist curators	Karp et al. (2017)

(continued)

Table 3.4 (continued)

S. No.	Metabolomics database	Information	References
13.	BRENDA (The Comprehensive Enzyme Information System)	Comprehensive enzyme information system, including FRENDA, AMENDA, DRENDA, and KENDA	Scheer et al. (2011)
14.	WikiPathways	Community resource for contributing and maintaining content dedicated to biological pathways	Pico et al. (2008)
15.	Metabolomics at Rothamsted (MeT-RO)	Large- and small-scale metabolomic analyses of any plant or microbial material	Rothamsted Ltd

are small regulators of gene expression in the numerous developmental and signalling pathways and are emerging as important post-transcriptional regulators that may regulate key plant genes responsible for stress tolerance. Plants combat environmental stresses by activating several gene regulatory pathways and studies with different model plants have revealed the role of these miRNAs in response to abiotic stress (Zhou et al. 2010). Plant exposed to abiotic stress causes over- or underexpression of certain miRNA and might even lead to the synthesis of new miRNAs to withstand stress (Khraiwesh et al. 2012). Several studies identified species- and clades-specific miRNA families associated with plant stress-regulated genes (Zhang et al. 2013). The functions of stress-responsive miRNAs can only be studied by understanding the regulatory interaction within the network (Jeong and Green 2013). Identification of a huge number of stress-responsive miRNAs might be helpful in developing new strategies to withstand stress, thereby improving the stress tolerance in plant. With the drastic improvement in genomic tools and methods, novel miRNAs in various plant species involved in abiotic stress response are increasing and are providing us with a better understanding of miRNAs-mediated gene regulation (Wang et al. 2014).

Sequence-based profiling along with computational analysis has played a key role in the identification of stress-responsive miRs. sRNA blot and RT-PCR analysis have played an equally important part in systematically confirming the profiling data (Jagadeeswaran et al. 2010). This has also enabled quantification of their effect on the genetic networks, such that many of the stress-regulated miRs have emerged as potential candidates for improving plant performance under stress. The development and integration of plant computational biology tools and approaches have added new functionalities and perspectives in the miR biology to make them relevant for genetic engineering programmes for enhancing abiotic stress tolerance. So far, three major strategies have been employed for the identification and expression profiling of stress-induced miRs:

- The first approach involves the classical experimental route that included direct cloning, genetic screening or expression profiling.
- The second method involved computational predictions from genomic or EST loci.
- The latest one employed a combo of both as it was based on the prediction of miRs from high-throughput sequencing (HTS) data.

Each of these was followed by experimental validations by northern analysis, PCRs or microarrays. In recent years, high-throughput sequencing and screening protocols have caused an exponential increase in a number of miRs, identified and functionally annotated from various plant species (Jagadeeswaran et al. 2010). The first biological database generated for miR was miRBase, which acts as an archive of miR sequences and annotations (Griffiths-Jones et al. 2008). With the future advancement of genomic tools and methods to identify novel miRNAs in various plant species, the number of miRNAs involved in abiotic stress response is increasing, thus providing us with a better understanding of miRNAs-mediated gene regulation during various abiotic stresses (Table 3.5). Some of the databases comprising miRNAs-related information are:

- *PASmiR:* This database is a complete repository for miRNA regulatory mechanisms involved in plant response to abiotic stresses for the plant stress physiology community. It is a literature-curated and web-accessible database and was developed to provide detailed, searchable descriptions of miRNA molecular regulation in different plant abiotic stresses. It currently includes data from ~200 published studies, representing 1038 regulatory relationships between 682 miRNAs and 35 abiotic stresses in 33 plant species (Zhang et al. 2013).
- *PmiRExAt:* It is a new online database resource that caters plant miRNA expression atlas. The web-based repository comprises of miRNA expression profile and query tool for 1859 wheat, 2330 rice and 283 maize miRNA. The database interface offers open and easy access to miRNA expression profile and helps in identifying tissue preferential, differential and constitutively expressing miRNAs (Table 3.5).
- *Plant MicroRNA Database (PMRD):* miRNA expression profiles are provided in this database, including rice oxidative stress-related microarray data and the published microarray data for poplar, *Arabidopsis*, tomato, maize and rice. The plant miRNA database integrates available plant miRNA data deposited in public databases, gathered from the recent literature, and data generated in-house (Zhang et al. 2010).
- *WMP:* It is a novel resource that provides data related to the expression of abiotic stress-responsive miRNAs in wheat. This database allows the query of small RNA libraries, including in silico predicted wheat miRNA sequences and the expression profiles of small RNAs identified from those libraries (Remita et al. 2016).

3.3 Role of Bioinformatics in Plant Disease Management

Omics studies focused on whole-genome analysis have unlocked a new era for biology in general and for agriculture in particular. Combination of bioinformatics and functional genomics globally has paved way towards a better understanding of plant-pathogen biological interaction which eventually leads to breaking thoughts in the promotion of plant resistance to pests (Koltai and Volpin 2003). Bioinformatics

Table 3.5 Major microRNA repositories and stress databases

S. No.	Databases	Information	References
1.	miRcheck	The tool requires the input of putative hairpin sequences and their secondary structures; the algorithm was first used for identifying conserved miRs in *Arabidopsis* and rice	Jones-Rhoades and Bartel (2004)
2.	PmiRExAt	Plant miRNA expression atlas database and web applications	Gurjar et al. (2016)
3.	UEA sRNA Workbench	It predicts miR from HTS data, trans-acting RNA prediction, the secondary structure of RNA sequences, expression patterns of sRNA loci, alignment of short reads to the genome	Moxon et al. (2008)
4.	TAPIR	Characterizes miR-target duplexes with large loops which are usually not detectable by traditional target prediction tools, fast and canonical FASTA local alignment programme and RNAhybrid for detection of miR-mRNA duplexes	Pearson (2004), Krüger and Rehmsmeier (2006), and Bonnet et al. (2010)
5.	CLCGenomics Workbench	Calculating abundances of sRNA libraries. Workbench provides an interactive visualization to the differential expression and statistical analysis of RNA-Seq and sRNA data	Matvienko et al. (2013)
6.	C-mii	Tool aligns known miRs from different plant species to the EST sequences of the query plant species using blast homology search and has a unique feature of predicting the secondary structures of the miR-target duplexes	Numnark et al. (2012)
7.	miRDeep-P	Specialized tool for identification of plant miR collection of PERL scripts that are used for prediction of novel miRs from deep sequencing data	Friedländer et al. (2008) and Yang and Li (2011)
8.	CleaveLand	The general pipeline, available as a combination of PERL scripts, for detecting miR-cleaved target transcripts from degradome datasets	Addo-Quaye et al. (2008)
9.	ARMOUR	Datasets of rice miRs from various deep sequencing datasets for examining the expression changes with respect to their targets, a valuable tool to biologists for selecting miRs for further functional studies	Tripathi et al. (2015)
10.	PASmiR	Database for the role of miRNA in response to plant abiotic stress	Zhang et al. (2013)
11.	PMRD – plant microRNA database	Plant miRNA database integrates available plant miRNA data deposited in public databases, gleaned from the recent literature, and data generated in-house	Zhang et al. (2010)
12.	WMP	Novel comprehensive wheat miRNA database, including related bioinformatics software	Remita et al. (2016)

has played a great role in plant disease management by understanding the molecular basis of the host-pathogen interaction (Koltai and Volpin 2003). Modern genomics tools, including applications of bioinformatics and functional genomics, allow scientists to interpret DNA sequence data and test hypotheses on a larger scale than previously possible (Anonymous 2005). From past few years, numerous components of the plant signalling system have also been identified that function downstream of the detection molecules such as the pathogen proteins that are used to suppress host defences and drive the infection process (so-called effector proteins) by using molecular biological technologies and genetics approaches (Anonymous 2005). Disease resistance is only one of the several traits under selection in a breeding programme. Thus, bioinformatics has to play an increasing role in integrating phenotypic and pedigree information for agronomic as well as resistance traits (Vassilev et al. 2006). Improved algorithms and increased computing power have made it possible to improve selection strategies as well as to model the epidemiology of pathogens (Michelmore 2003). Some of the key roles of bioinformatics for plant improvements has been enlisted by Vassilev et al. (2005): submitting all sequence data information generated from experimentation into the public domain, through repositories; providing rational annotation of genes, proteins and phenotypes; elaborating relationships both within the plants' data and between plants and other organisms; providing data including information on mutations, markers, maps, functional discoveries; and others.

From past few years, there have been many technological advances in the understanding of plant-pathogen interaction. Omics techniques (genomics, proteomics and transcriptomics) have provided a great opportunity to explore plant-pathogen interactions from a system's perspective and studies on protein-protein interactions (PPIs) between plants and pathogens (Delaunois et al. 2014). Identification of the molecular components as well as the corresponding pathways has provided a relatively clear understanding of the plant immune system. In particular, the study of plant-pathogen interactions has also been stimulated by the emergence of various omics techniques, such as genomics, proteomics and transcriptomics (Schulze et al. 2015). With the availability of massive amounts of data generated from high-throughput omics techniques, network interactions have become a powerful approach to further decipher the molecular mechanisms of plant-pathogen interactions through network biology:

- Genomics is particularly important, and with the rapid development of next-generation sequencing (NGS) technique, numerous plant and pathogen genomes have been fully sequenced.
- Proteomics is a key technique for the analysis of the proteins involved in plant-pathogen interactions (Delaunois et al. 2014).
- DNA microarray and RNA sequencing are two key transcriptomics techniques for acquiring the expression profile of genes on a large scale. Transcriptomics is also important to investigate plant-pathogen interactions and has been employed to learn how plants respond to the pathogen invasion and how pathogens counter the plant defence at the transcript level (Schulze et al. 2015).

Genomic approaches always have a significant impact on efforts to improve plant diseases by increasing the definition of and access to gene pools available for crop improvement (Vassilev et al. 2005). Such an approach to identify key genes and understand their function will result in a quantum leap in plant improvement. Moreover, the ability to examine gene expression will allow us to understand how plants respond to and interact with the physical environment and management practices (Vassilev et al. 2006). This approach will involve the detailed characterization of the many genes that confer resistance, as well as technologies for the precise manipulation and deployment of resistance genes. Plant-pathogen interactions are sophisticated and dynamic in the continually evolving competition between pathogens and plants. Thus genomic studies on pathogens are providing an understanding of the molecular basis of specificity and the opportunity to select targets for more durable resistance (Michelmore 2003). This understanding is fundamental to allow efficient exploitation of plants as biological resources in the development of new cultivars with improved quality and reduced economic, pathogen and abiotic stress resistance and is also vital for the development of new plant diagnostic tools (Vassilev et al. 2006). When plants respond to biotic stress, a series of biological processes rather than a single gene or protein will be changed. Therefore, it is necessary to explore plant-pathogen interactions from a systems perspective (e.g. network level (Mine et al. 2014)). Bioinformatics thus plays several roles in breeding for disease resistance and is important for acquiring and organizing large amounts of information. Some of the databases/repositories (Table 3.6) of plant-pathogen interactions are:

- *PHI-base: A new database for pathogen-host interactions.* It is designed for hosting any type of pathogen-host interaction, and its focus is on genes with functions that have been experimentally verified. These genes are compiled and curated in a way that can be used to bridge the genotype-phenotype gap underlying the interactions between hosts and pathogens (Winnenburg et al. 2006). The mission of PHI-base is to provide expertly curated molecular and biological information on genes proven to affect the outcome of pathogen-host interactions.
- *PathoPlant: A database on plant-pathogen interactions.* This is a relational database to display relevant components and reactions involved in signal transduction related to plant-pathogen interactions. On the organism level, the tables 'plant', 'pathogen' and 'interaction' are used to describe incompatible interactions between plants and pathogens or diseases (Bülow et al. 2004).

3.4 Conclusion and Future Prospects

Bioinformatics is an exclusive approach capable of exploiting and sharing a large amount of omics data. This approach has given a more holistic view of the molecular response in plants when exposed to biotic and abiotic stress, and the integration of various omics studies has revealed a new zone of interactions and regulation (Fig. 3.2). This system biology approach has enabled the identification,

Table 3.6 Abiotic stress databases

S. No	Databases	Information	References
1.	PHI-base (Pathogen-Host Interactions)	First online resource devoted to the identification and presentation of information on fungal and oomycete pathogenicity genes and their host interactions	Winnenburg et al. (2006)
2.	PathoPlant (Plant-Pathogen Interactions)	A new database that combines information of specific plant–pathogen interactions on organism level and data about signal transduction on molecular level related with plant pathogenesis	Bülow et al. (2004)
3.	Fungal Plant Pathogen Database	An internet-based database that crosslinks the digitized genotypic and phenotypic information of individual pathogens at both the species and population levels may allow us to effectively address these problems by coordinating the generation of data and its subsequent archiving	Kang et al. (2002)
4.	Phytopathogenic Fungi and Oomycete EST Database of COGEME	Provides sequences of expressed sequence tags (ESTs) and consequences (cluster assembled ESTs) from 15 plant pathogenic species	Soanes et al. (2002)
5.	fuGIMS database	Integrate functional and sequence information from several plants and animal pathogenic fungi with similar information from *Saccharomyces cerevisiae* available from the GIMS database	Cornell et al. (2003)
6.	DRASTIC Insight	Collates signal transduction information between plants, pathogens, and the environment, including both biotic and abiotic influences on plant disease resistance at the molecular level	Newton et al. (2002)

characterization and functional analysis of plant genes that determine plant's response to various biotic and abiotic factors and understanding of the plant stress interaction. However, so many efforts are still required for detailed analysis of the omic modulation induced by abiotic stress and its interacting partners. This requires the development of reliable and rigorous techniques for firm characterization of the spatiotemporal regulation of omics under stress conditions. The three main domains that must be addressed to take full advantage of systems biology are the development of omics technology, integration of data in a usable format and analysis of data within the domain of bioinformatics. Thus, the perspective of computational/system biology needs to be tapped for performing an extensive analysis among agriculturally important crops for improving crop tolerance to environmental stress. The current surge of affordable omics data encourages researchers to create improved, more integrated and easily accessible plant stress pathway databases. Despite the drawbacks, there is no doubt that bioinformatics is a field that holds great potential for transforming biological research in the coming decades. The expansion and integration of bioinformatics tools and approaches will certainly add new functionalities and perspectives in the stress biology to make them applicable for genetic engineering programmes for enhancing stress tolerance.

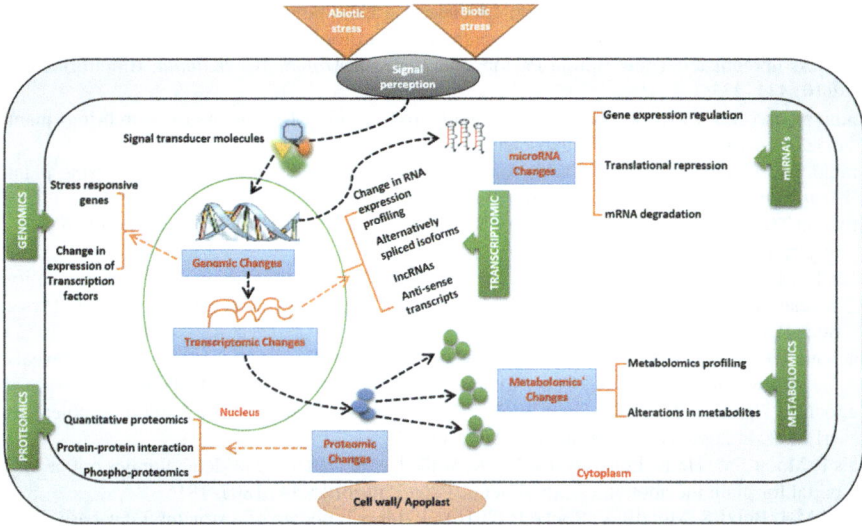

Fig. 3.2 Pathway depicting molecular effects of abiotic and biotic stress at genomic, transcriptomic, metabolomic and microRNA levels inside the plant cell. Both biotic and abiotic stresses have to be first sensed by the plant cell, and then the information is transduced to the appropriate downstream-located pathway(s). Sensors as well as signal transducers might be shared by both types of stressors. After the signal is perceived by the sensors at the cell wall, it transduces the signal towards the nucleus, where modifications occur at genome level, resulting in activation of stress-responsive genes and certain transcription factors (genomics). This activation can occur at the transcript level (transcriptomics), wherein a change in RNA expression levels, alternatively spliced forms of RNA and antisense transcripts occur. Few regulatory small RNAs (microRNA) are also synthesized against a specific type of stress conditions (miRNAs). Alterations, which have been initiated at the cell wall, propagate to the cytosol, in form of proteins which can be identified by quantitative proteomics, protein-protein interaction and phosphor-proteomics (proteomics). After stress signalling, the defence system comes into rescue where large amounts of secondary metabolites are synthesized (metabolomics); these metabolites can be analysed by metabolomic profiling and change in the type of metabolites synthesized

References

Adams MD, Soares MB, Kerlavage AR, Fields C, Venter JC (1993) Rapid cDNA sequencing (expressed sequence tags) from a directionally cloned human infant brain cDNA library. Nat Genet 4:373–380

Addo-Quaye C, Miller W, Axtell MJ (2008) CleaveLand: a pipeline for using degradome data to find cleaved small RNA targets. Bioinformatics 25(1):130–131

Agarwal P, Parida SK, Mahto A et al (2014) Expanding frontiers in plant transcriptomics in aid of functional genomics and molecular breeding. Biotechnol J 9:1480–1492

Akiyama K, Chikayama E, Yuasa H, Shimada Y, Tohge T, Shinozaki K, Hirai MY, Sakurai T, Kikuchi J, Saito K (2008) PRIMe: a web site that assembles tools for metabolomics and transcriptomics. Silicon Biol 8(3):339–345

Alter S, Bader KC, Spannagl M, Wang Y, Bauer E, Schön CC, Mayer KFX (2015) DroughtDB: an expert-curated compilation of plant drought stress genes and their homologs in nine species. J Biol Database Curation 2015:bav046

Ambika S, Susan Mary Varghese SM, Shameer K, Udayakumar M, Sowdhamini R (2008) STIF: Hidden Markov Model-based search algorithm for the recognition of binding sites of Stress-upregulated Transcription Factors and genes in *Arabidopsis thaliana*. Bioinformation 2(10):431–437

Anonymous (2005) Pseudomonas versus Arabidopsis: models for genomic research into plant disease resistance. www.actionbioscience.org

Arabidopsis Genome Initiative (2000) Analysis of the genome sequence of the flowering plant *Arabidopsis thaliana*. Nature 408(6814):796

Attwood TK, Beck ME, Bleasby AJ, Parry-Smith DJ (1994) PRINTS–a database of protein motif fingerprints. Nucleic Acids Res 22(17):3590–3596

Badjakov I, Kondakova V, Atanassov A (2012) In: Benkeblia N (ed) Current view on fruit quality in relation to human health in sustainable agriculture and new biotechnologies. CRC Press, Boca Raton, pp 303–319

Baerenfaller K, Grossmann J, Grobei MA et al (2008) Genome-scale proteomics reveals *Arabidopsis thaliana* gene models and proteome dynamics. Science 320:938–941

Bairoch A, Apweiler R (2000) The SWISS-PROT protein sequence database and its supplement TrEMBL in 2000. Nucleic Acids Res 28(1):45–48

Bais P, Moon SM, He K, Leitao R, Dreher K, Walk T et al (2015) PlantMetabolomics.org: a web portal for plant metabolomics experiments. Plant Physiol 152(4):1807–1816

Basse M-J, Betzi S, Morelli X, Roche P (2016) 2P2Idb v2: update of a structural database dedicated to orthosteric modulation of protein–protein interactions. Database 2016:baw007

Bino RJ, Hall RD, Fiehn O, Kopka J, Saito K, Draper J, Nikolau BJ, Mendes P, Roessner-Tunali U, Beale MH et al (2004) Potential of metabolomics as a functional genomics tool. Trends Plant Sci 9:418–425

Bolser D, Staines DM, Pritchard E, Kersey P (2016) Ensembl plants: integrating tools for visualizing, mining, and analyzing plant genomics data. In: Edwards D (ed) Plant bioinformatics, Methods in molecular biology, vol 1374. Humana Press, New York, NY

Bonnet E, He Y, Billiau K, Van de Peer Y (2010) TAPIR, a web server for the prediction of plant microRNA targets, including target mimics. Bioinformatics 26:1566–1568

Borkotoky S, Saravanan V, Jaiswal A, et al. (2013) The Arabidopsis stress responsive gene database. Int J Plant Genom 2013:949564

Broughton WJ, Hernández G, Blair M et al (2003) Beans (Phaseolus spp.)—model food legumes. Plant Soil 252:55–128

Brown JW, Shaw PJ, Shaw P, Marshall DF (2005) Arabidopsis nucleolar protein database (AtNoPDB). Nucleic Acids Res 33:D633–D636

Bülow L, Schindler M, Choi C, Hehl R (2004) PathoPlant: a database on plant-pathogen interactions. In Silico Biol 4(4):529–536

Castellana NE, Payne SH, Shen Z et al (2008) Discovery and revision of Arabidopsis genes by proteogenomics. Proc Natl Acad Sci U S A 105:21034–21038

Castillo-Peinado LS, de Castro ML (2016) Present and foreseeable future of metabolomics in forensic analysis. Anal Chim Acta 925:1–5

Chawla K, Barah P, Kuiper M, Bones AM (2011) Systems biology: a promising tool to study abiotic stress responses. Omics Plant Abiotic Stress Tolerance 10:163–172

Chen X, Qi X, Duan LX (2015) Overview. In: Plant metabolomics. Springer, Netherlands, p 1–24

Coman D, Gruissem W, Hennig L (2013) Transcript profiling in Arabidopsis with genome tiling microarrays. In: Tiling arrays: methods and protocols. Humana Press, Totowa, pp 35–49

Cook D, Fowler S, Fiehn O, Thomashow MF (2004) A prominent role for the CBF cold response pathway in configuring the low-temperature metabolome of Arabidopsis. Proc Natl Acad Sci U S A 101:15243–15248

Cooke IR, Jones D, Bowen JK et al (2014) Proteogenomic analysis of the Venturia pirina (Pear Scab Fungus) secretome reveals potential effectors. J Proteome Res 13:3635–3644

Cornell M, Paton NW, Hedeler C, Kirby P, Delneri D, Hayes A, Oliver SG (2003) GIMS: an integrated data storage and analysis environment for genomic and functional data. Yeast 20(15):1291–1306

Cushman JC, Bohnert HJ (2000) Genomic approaches to plant stress tolerance. Curr Opin Plant Biol 3:117–124

Dash S, Van Hemert J, Hong L, Wise RP, Dickerson JA (2012) PLEXdb: gene expression resources for plants and plant pathogens. Nucleic Acids Res 40:D1194–D1201

Dash S, Campbell JD, Cannon EK, Cleary AM, Huang W, Kalberer SR, Karingula V, Rice AG, Singh J, Umale PE, Weeks NT, Wilkey AP, Farmer AD, Cannon SB (2016) Legume information system (LegumeInfo.org): a key component of a set of federated data resources for the legume family. Nucleic Acids Res 44:D1181–D1188

De Cremer K, Mathys J, Vos C et al (2013) RNAseq-based transcriptome analysis of *Lactuca sativa* infected by the fungal necrotroph Botrytis cinerea. Plant Cell Environ 36:1992–2007

Delaunois B, Jeandet P, Clément C, Baillieul F, Dorey S, Cordelier S (2014) Uncovering plant–pathogen crosstalk through apoplastic proteomic studies. Front Plant Sci 5:249

Delmotte N, Knief C, Chaffron S et al (2009) Community proteogenomics reveals insights into the physiology of phyllosphere bacteria. Proc Natl Acad Sci U S A 106:16428–16433

Dong Q, Schlueter SD, Brendel V (2004) PlantGDB, plant genome database and analysis tools. Nucleic Acids Res 32:D354–D359

Dubois A, Carrere S, Raymond O, Pouvreau B, Cottret L, Roccia A, Onesto JP, Sakr S, Atanassova R, Baudino S, Foucher F, Le Bris M, Gouzy J, Bendahmane M (2012) Transcriptome database resource and gene expression atlas for the rose. BMC Genomics 13:638

Duque AS, de Almeida AM, da Silva AB, da Silva JM, et al (2013) Abiotic stress—plant responses and applications in agriculture. InTech, Chapter 3, p 40–101

Durek P, Schmidt R, Heazlewood JL, Jones A, Maclean D, Nagel A, Kersten B, Schulze WX (2010) PhosPhAt: the *Arabidopsis thaliana* phosphorylation site database. An update. Nucleic Acids Res 38:D828–D834

Eldem V, Okay S, Ünver T (2013) Plant microRNAs: new players in functional genomics. Turk J Agric For 37:1–21

El-Metwally S, Ouda OM, Helmy M (2014a) Next generation sequencing technologies and challenges in sequence assembly, 1st edn. Springer, ISBN: 978-1-4939-0714-4

El-Metwally S, Ouda OM, Helmy M (2014b) First- and next-generations sequencing methods. Next gener seq technol Challenges seq assem. Springer, New York, pp 29–36

El-Metwally S, Ouda OM, Helmy M (2014c) New horizons in next-generation sequencing. Next gener seq technol Challenges seq assem. Springer, New York, pp 51–59

Ernst M, Silva DB, Silva RR, Vêncio RZ, Lopes NP (2014) Mass spectrometry in plant metabolomics strategies: from analytical platforms to data acquisition and processing. Nat Prod Rep 31(6):784–806

Esposito A, Colantuono C, Ruggieri V, Chiusano ML (2016) Bioinformatics for agriculture in the next-generation sequencing era. Chem Biol Technol Agric 3(1):9

Fernie AR, Schauer N (2009) Metabolomics-assisted breeding: a viable option for crop improvement? Trends Genet 25(1):39–48

Feuillet C, Leach JE, Rogers J, Schnable PS, Eversole K (2011) Crop genome sequencing: lessons and rationales. Trends Plant Sci 16(2):77–88

Finn RD, Bateman A, Clements J, Coggill P, Eberhardt RY, Eddy SR, Heger A, Hetherington K, Holm L, Mistry J, Sonnhammer ELL, Tate J, Punta M (2013) Pfam: the protein families database. Nucleic Acids Res 42(D1):D222–D230

Friedländer MR, Chen W, Adamidi C, Maaskola J, Einspanier R, Knespel S et al (2008) Discovering microRNAs from deep sequencing data using miRDeep. Nat Biotechnol 26:407–415

Futamura N, Totoki Y, Toyoda A, Igasaki T, Nanjo T, Seki M et al (2008) Characterization of expressed sequence tags from a full-length enriched cDNA library of *Cryptomeria japonica* male strobili. BMC Genomics 9:383

Gao J, Agrawal GK, Thelen JJ, Xu D (2009) Helmy Nucleic Acids Res 37:D960–D962

Ghosh D, Xu J (2014) Abiotic stress responses in plant roots: a proteomics perspective. Front Plant Sci 5:6

Gomez-Casati DF, Zanor MI, Busi MV (2013) Metabolomics in plants and humans: applications in the prevention and diagnosis of diseases. Biomed Res Int 2013:1–11

Govind G, Harshavardhan VT, Patricia JK et al (2009) Identification and functional validation of a unique set of drought induced genes preferentially expressed in response to gradual water stress in peanut. Mol Gen Genomics 281:607

Govindaraj M, Vetriventhan M, Srinivasan M (2015) Importance of genetic diversity assessment in crop plants and its recent advances: an overview of its analytical perspectives. Genet Res Int 2015:431487

Griffiths-Jones S, Saini HK, vanDongen S, Enright AJ (2008) miRBase: tools for microRNA genomics. Nucleic Acids Res 36:D154–D158

Gupta P, Naithani S, Tello-Ruiz MK, Chougule K, D'Eustachio P, Fabregat A et al (2016) Gramene database: navigating plant comparative genomics resources. Curr Plant Biol 8:10–15

Gurjar AKS, Singh Panwar A, Gupta R, Mantri SS (2016) PmiRExAt: plant miRNA expression atlas database and web applications. Database 2016:baw060

Hammami R, Ben Hamida J, Vergoten G, Fliss I (2009) PhytAMP: a database dedicated to antimicrobial plant peptides. Nucleic Acids Res 37:D963–D968

Harbers M, Carninci P (2005) Tag-based approaches for transcriptome research and genome annotation. Nat Methods 2:495–502

Heazlewood JL, Durek P, Hummel J, Selbig J, Weckwerth W, Walther D, Schulze WX (2008) PhosPhAt: a database of phosphorylation sites in Arabidopsis thaliana and a plant-specific phosphorylation site predictor. Nucleic Acids Res 36:D1015–D1021

Helmy M, Tomita M, Ishihama Y (2011) OryzaPG-DB: rice proteome database based on shotgun proteogenomics. BMC Plant Biol 11:63

Helmy M, Sugiyama N, Tomita M, Ishihama Y (2012a) Mass spectrum sequential subtraction speeds up searching large peptide MS/MS spectra datasets against large nucleotide databases for proteogenomics. Cell Mech 17:633–644

Helmy M, Sugiyama N, Tomita M, Ishihama Y (2012b) The rice proteogenomics database oryza PG-DB: development, expansion, and new features. Front Plant Sci 3:65

Hopff D, Wienkoop S, Lüthje S (2013) The plasma membrane proteome of maize roots grown under low and high iron conditions. J Proteome 91:605–618

Hunter S, Apweiler R, Attwood TK, Bairoch A, Bateman A, Binns D, Bork P, Das U, Daugherty L (2009) InterPro: the integrative protein signature database. Nucleic Acids Res 37:D211–D215

Hulo N (2006) The PROSITE database. Nucleic Acids Res 34(90001):D227–D230

International Rice Genome Sequencing Project (2005) The map-based sequence of the rice genome. Nature 436:793–800

Iijima Y, Nakamura Y, Ogata Y, Tanaka K'i, Sakurai N, Suda K, Suzuki T, Suzuki H, Okazaki K, Kitayama M, Kanaya S, Aoki K, Shibata D (2008) Metabolite annotations based on the integration of mass spectral information. Plant J 54(5):949–962

Jagadeeswaran G, Zheng Y, Sumathipala N, Jiang H, Arrese EL, Soulages JL et al (2010) Deep sequencing of small RNA libraries reveals dynamic regulation of conserved and novel microRNAs and microRNA-stars during silkworm development. BMC Genomics 11:52

Jayashree B, Crouch JH, Prasad PVNS, Hoising-ton D (2006) A database of annotated tentative orthologs from crop abiotic stress transcripts. Bioinformation 1:225–227

Jeong DH, Green PJ (2013) The role of rice microRNAs in abiotic stress responses. J Plant Biol 56:187–197

Jones-Rhoades MW, Bartel DP (2004) Computational identification of plant MicroRNAs and their targets, including a stress-induced miRNA. Mol Cell 14(6):787–799

Jogaiah S, Govind SR, Tran L-SP (2013) Systems biology-based approaches toward understanding drought tolerance in food crops. Crit Rev Biotechnol 33:23–39

Jorrín-Novo JV, Maldonado AM, Echevarría-Zomeño S, Valledor L, Castillejo MA, Curto M, Valero J, Sghaier B, Donoso G, Redondo I (2009) Plant proteomics update (2007–2008): second-generation proteomic techniques, an appropriate experimental design, and data analysis to fulfill MIAPE standards, increase plant proteome coverage and expand biological knowledge. J Proteome 72(3):285–314

Joshi R, Karan R, Singla-Pareek SL, Pareek A (2012) Microarray technology. In: Gupta AK, Pareek A, Gupta SM (eds) Biotechnology in medicine and agriculture: principles and practices. IK International Publishing House Pvt. Ltd., New Delhi, pp 273–296

Kanehisa M, Goto S (2000) KEGG: kyoto encyclopedia of genes and genomes. Nucleic Acids Res 28(1):27–30

Kang S, Ayers JE, Dewolf ED, Geiser DM, Kuldau G, Moorman GW, Mullins E, Uddin W, Correll JC, Deckert G, Lee YH, Lee YW, Martin FN, Subbarao K (2002) The internet-based fungal pathogen database: a proposed model. Phytopathology 92(3):232–236

Karp PD, Billington R, Caspi R, Fulcher CA, et al. (2017) The BioCyc collection of microbial genomes and metabolic pathways. Brief Bioinform. https://doi.org/10.1093/bib/bbx085

Kawahara Y, de la Bastide M, Hamilton JP, Kanamori H, McCombie et al (2013) Improvement of the *Oryza sativa* Nipponbare reference genome using next generation sequence and optical map data. Rice 6:4

Kawahara Y, Oono Y, Wakimoto H, Ogata J (2016) TENOR: database for comprehensive mRNA-Seq experiments in rice. Plant Cell Physiol 57(1):e7

Khraiwesh B, Zhu JK, Zhu J (2012) Role of miRNAs and siRNAs in biotic and abiotic stress responses of plants. Biochim Biophys Acta 1819:137–148

Kim HS, Mittenthal JE, Caetano-Anolles G (2006) MANET:tracing evolution of protein architecture in metabolic networks. BMC Bioinforma 7:351

King ZA, Lu JS, Dräger A, Miller PC, Federowicz S, Lerman JA, Ebrahim A, Palsson BO, Lewis NE (2016) BiGG Models: a platform for integrating, standardizing, and sharing genome-scale models. Nucleic Acids Res 44(D1):D515–D522

Kinjo AR, Bekker GJ, Suzuki H, Tsuchiya Y, Kawabata T, Ikegawa Y, Nakamura H (2017) Protein Data Bank Japan (PDBj): updated user interfaces, resource description framework, analysis tools for large structures. Nucleic Acids Res 45:D282–D288

Kissoudis C, van de Wiel C, Visser RGF, van der Linden G (2014) Enhancing crop resilience to combined abiotic and biotic stress through the dissection of physiological and molecular cross-talk. Front Plant Sci 5:e207

Koltai H, Volpin H (2003) Agricultural genomics: an approach to plant protection. Eur J Plant Pathol 109:101–108

Kopka J, Schauer N, Krueger S, Birkemeyer C, Usadel B, Bergmuller E, Dormann P, Weckwerth W, Gibon Y, Stitt M, Willmitzer L, Fernie AR, Steinhauser D (2005) GMD@CSB.DB: the Golm metabolome database. Bioinformatics 21(8):1635–1638

Krüger J, Rehmsmeier M (2006) RNAhybrid: microRNA target prediction easy, fast and flexible. Nucleic Acids Res 34:W451–W454

Kudo T, Terashima S, Takaki Y, Tomita K et al (2017) PlantExpress: a database integrating OryzaExpress and ArthaExpress for single-species and cross-species gene expression network analyses with microarray-based transcriptome data. Plant Cell Physiol 58(1):e1

Lee TH, Kim YK, Pham THM, Song SI et al (2009) Correlating gene expression from transcriptome profiling, and its application to the analysis of coexpressed genes in rice. Plant Physiol 151(1):16–33

Li JR, Liu CC, Sun CH, Chen YT (2018) Plant stress RNA-seq nexus: a stress-specific transcriptome database in plant cells. BMC Genomics 19:966

Liu B, Zhang N, Zhao S et al (2015) Proteomic changes during tuber dormancy release process revealed by iTRAQ quantitative proteomics in potato. Plant Physiol Biochem 86:181–190

Makita Y, Shimada S, Kawashima M, Kondou-Kuriyama T, Toyoda T, Matsui M (2015) MOROKOSHI: transcriptome database in *Sorghum bicolor*. Plant Cell Physiol 56:1

Maruyama K, Takeda M, Kidokoro S, Yamada K, Sakuma Y, Urano K, Fujita M, Yoshiwara K, Matsukura S, Morishita Y, Sasaki R (2009) Metabolic pathways involved in cold acclimation identified by integrated analysis of metabolites and transcripts regulated by DREB1A and DREB2A. Plant Physiol 150(4):1972–1980

Masoudi-Nejad A, Goto S, Jauregui R et al (2007) EGENES: transcriptome-based plant database of genes with metabolic pathway information and expressed sequence tag indices in KEGG. Plant Physiol 144(2):857–866

Matsuda F, Yonekura-Sakakibara K, Niida R, Kuromori T, Shinozaki K, Saito K (2009) MS/MS spectral tag based annotation of non-targeted profile of plant secondary metabolites. Plant J 57(3):555–577

Matsui A, Ishida J, Morosawa T, Mochizuki Y, Kaminuma E, Endo TA, Okamoto M, Nambara E, Nakajima M, Kawashima M, Satou M (2008) Arabidopsis transcriptome analysis under drought, cold, high-salinity and ABA treatment conditions using a tiling array. Plant Cell Physiol 49(8):1135–1149

Matvienko M, Kozik A, Froenicke L, Lavelle D, Martineau B, Perroud B et al (2013) Consequences of normalizing transcriptomic and genomic libraries of plant genomes using a duplex-specific nuclease and tetramethylammonium chloride. PLoS One 8(2):e55913

McCarthy FM, Wang N, Bryce Magee G, Nanduri B, Lawrence ML, Camon EB, Barrell DG, Hill DP, Dolan ME, Paul Williams W, Luthe DS, Bridges SM, Burgess SC (2006) AgBase: a functional genomics resource for agriculture. BMC Genomics 7(1):229

Michelmore RW (2003) The impact zone: genomics and breeding for durable disease resistance. Curr Opin Plant Biol 6:397–404

Mine A, Sato M, Tsuda K (2014) Toward a systems understanding of plant–microbe interactions. Front Plant Sci 5:423

Mir S, Alhroub Y, Anyango S, Armstrong DR, Berrisford JM, Clark AR et al (2018) PDBe: towards reusable data delivery infrastructure at protein data bank in Europe. Nucleic Acids Res 46:D486–D492

Moco S, Bino RJ, Vorst O, Verhoeven HA, de Groot J, van Beek TA, Vervoort J, de Vos CHR (2006) A liquid chromatography-mass spectrometry-based metabolome database for tomato. Plant Physiol 141(4):1205–1218

Mousavi SA, Pouya FM, Ghaffari MR, Mirzaei M, Ghaffari A, Alikhani M, Ghareyazie M, Komatsu S, Haynes PA, Salekdeh GH (2016) PlantPReS: a database for plant proteome response to stress. J Proteome 143:69–72

Moxon S, Schwach F, Dalmay T, Maclean D, Studholme DJ, Moulton V (2008) A toolkit for analysing large-scale plant small RNA datasets. Bioinformatics 24:2252–2253

Murzin AG, Brenner SE, Hubbard T, Chothia C (1995) SCOP: a structural classification of proteins database for the investigation of sequences and structures. J Mol Biol 247:536–540

Naika M, Shameer K, Mathew OK, Gowda R, Sowdhamini R (2013) STIFDB2: an updated version of plant stress-responsive transcription factor database with additional stress signals, stress-responsive transcription factor binding sites and stress-responsive genes in Arabidopsis and rice. Plant Cell Physiol 54:1–15

Nakagami H, Sugiyama N, Ishihama Y, Shirasu K (2012) Shotguns in the front line: phosphoproteomics in plants. Plant Cell Physiol 53:118–124

Nakaya A, Ichihara H, Asamizu E, Shirasawa S, Nakamura Y, Tabata S, Hirakawa H (2017) Plant genomics databases. Methods in molecular biology, vol 1533. Humana Press, New York, pp 45–77

Narsai R, Ivanova A, Ng S, Whelan J (2010) Defining reference genes in Oryza sativa using organ, development, biotic and abiotic transcriptome datasets. BMC Plant Biol 10:56

Newton AC, Lyon GD, Marshall B (2002) DRASTIC: a database resource for analysis of signal transduction in cells. BSPP Newsl 42:36–37

Numnark S, Mhuantong W, Ingsriswang S, Wichadakul D (2012) C-mii: a tool for plant miRNA and target identification. BMC Genomics 13:S16

Osuna-Cruz CM, Paytuvi-Gallart A, Di Donato A, Sundesha V, Andolfo G, Aiese Cigliano R, Sanseverino W, Ercolano MR (2018) PRGdb 3.0: a comprehensive platform for prediction and analysis of plant disease resistance genes. Nucleic Acids Res 46:D1197–D1201

Pearson W (2004) Finding protein and nucleotide similarities with FASTA. Curr Protoc Bioinforma Chapter 3:Unit3.9

Pico AR, Kelder T, van Iersel MP, Hanspers K, Conklin BR, Evelo C (2008) WikiPathways: pathway editing for the people. PLoS Biol 6(7):e184

Prabha R, Ghosh I, Singh DP (2011) Plant stress gene database: a collection of plant genes responding to stress condition. ARPN J Sci Tech 1(1):28–31

Priya P, Jain M (2013) RiceSRTFDB: a database of rice transcription factors containing comprehensive expression, cis-regulatory element and mutant information to facilitate gene function analysis. J Database Curation Bat027:1–7

Ramegowda V, Senthil-Kumar M (2015) The interactive effects of simultaneous biotic and abiotic stresses on plants: mechanistic understanding from drought and pathogen combination. J Plant Physiol 176:47–54

Remita AM, Lord E, Agharbaoui Z, Leclercq M et al (2016) WMP: a novel comprehensive wheat miRNA database, including related bioinformatics software. Curr Plant Biol 7(8):31–33

Rhee SY, Beavis W, Berardini TZ, Chen G, Dixon D, Doyle A et al (2003) The Arabidopsis Information Resource (TAIR): a model organism database providing a centralized, curated gateway to Arabidopsis biology, research materials and community. Nucleic Acids Res 31(1):224

Rocha I, Förster J, Nielsen J (2008) Design and application of genome-scale reconstructed metabolic models. Microbi Gene Essentiality: Protoc Bioinforma 416:409–431

Rose PW, Prlić A, Altunkaya A, Bi C, Bradley AR, Christie CH et al (2017) The RCSB protein data Bank: integrative view of protein, gene and 3D structural information. Nucleic Acids Res 45:D271–D281

Sakurai T, Satou M, Akiyama K, Iida K, Seki M, Kuromori T et al (2005) RARGE: a large-scale database of RIKEN Arabidopsis resources ranging from transcriptome to phenome. Nucl Acids Res 33:D647–D650

Scheer M, Grote A, Chang A, Schomburg I, Munaretto C, Rother M, Söhngen C, Stelzer M, Thiele J, Schomburg D (2011) BRENDA, the enzyme information system in 2011. NucleicAcids Res 39:D670–D676

Schulze S, Henkel SG, Driesch D, Guthke R, Linde J (2015) Computational prediction of molecular pathogen–host interactions based on dual transcriptome data. Front Microbiol 6:65

Shafi A, Dogra V, Gill T, Ahuja PS, Sreenivasulu Y (2014) Simultaneous over-expression of PaSOD and RaAPX in transgenic *Arabidopsis thaliana* confers cold stress tolerance through increase in vascular lignifications. PLoS One 9:e110302

Shafi A, Gill T, Sreenivasulu Y, Kumar S, Ahuja PS, Singh AK (2015a) Improved callus induction, shoot regeneration, and salt stress tolerance in Arabidopsis overexpressing superoxide dismutase from *Potentilla atrosanguinea*. Protoplasma 252:41–51

Shafi A, Chauhan R, Gill T, Swarnkar MK, Sreenivasulu Y, Kumar S, Kumar N, Shankar R, Ahuja PS, Singh AK (2015b) Expression of SOD and APX genes positively regulates secondary cell wall biosynthesis and promotes plant growth and yield in Arabidopsis under salt stress. Plant Mol Biol 87:615–631

Shafi A, Pal AK, Sharma V, Kalia S, Kumar S, Ahuja PS, Singh AK (2017) Transgenic potato plants overexpressing SOD and APX exhibit enhanced lignification and starch biosynthesis with improved salt stress tolerance. Plant Mol Biol Rep 35:504–518

Shameer K, Ambika S, Varghese SM, Karaba N, Udayakumar M, Sowdhamini R (2009) STIFDB-Arabidopsis stress responsive transcription factor dataBase. Int J Plant Genomics 2009:583429

Shankar A, Singh A, Kanwar P et al (2013) Gene expression analysis of rice seedling under potassium deprivation reveals major changes in metabolism and signaling components. PLoS One 8:e70321

Singh VK, Singh AK, Chand R, Kushwaha C (2011) Role of bioinformatics in agriculture and sustainable development. Int J Bioinform Res 3(2):221–226

Singh B, Bohra A, Mishra S, Joshi R, Pandey S (2015) Embracing new-generation 'omics' tools to improve drought tolerance in cereal and food-legume crops. Biol Plant 59(3):413–428

Smita S, Lenka SK, Katiyar A, Jaiswal P, Preece J, Bansal KC (2011) QlicRice: a web interface for abiotic stress responsive QTL and loci interaction channels in rice. Database 1:1–9

Soanes DM, Skinner W, Keon J, Hargreaves J, Talbot NJ (2002) Genomics of phytopathogenic fungi and the development of bioinformatic resources. Mol Plant-Microbe Interact 15(5):421–427

Soga T, Ueno Y, Naraoka H et al (2002) Simultaneous determination of anionic intermediates for Bacillus subtilis metabolic pathways by capillary electrophoresis electrospray ionization mass spectrometry. Anal Chem 74:2233–2239

Somerville C, Dangl J (2000) Genomics. Plant biology in 2010. Science 290:2077–2078

Spannagl M, Nussbaumer T, Bader KC, Martis MM, Seidel M, Kugler KG, Gundlach H, Mayer KFX (2016) PGSB PlantsDB: updates to the database framework for comparative plant genome research. Nucleic Acids Res 44:D1141–D1147

Sun Q, Zybailov B, Majeran W, Friso G, Olinares PD, van Wijk KJ (2009) PPDB, the Plant Proteomics Database at Cornell. Nucleic Acids Res 37:D969–D974

Suzuki N, Rivero RM, Shulaev V, Blumwald E, Mittler R (2014) Abiotic and biotic stress combinations. New Phytol 203:32–43

Swarbreck D, Wilks C, Lamesch P et al (2008) The Arabidopsis information resource (TAIR): gene structure and function annotation. Nucleic Acids Res 36:D1009–D1014

Tardieu F, Tuberosa R (2010) Dissection and modelling of abiotic stress tolerance in plants. Curr Opin Plant Biol 13:206–212

Tripathi A, Goswami K, Mishra NS (2015) Role of bioinformatics in establishing microRNAs as modulators of abiotic stress responses: the new revolution. Front Physiol 26:286

Tyers M, Mann M (2003) From genomics to proteomics. Nature 422:193–197

Ueno S, Nakamura Y, Kobayashi M, Terashima S (2018) TodoFirGene: developing transcriptome resources for genetic analysis of Abies sachalinensis. Plant Cell Physiol 59(6):1276–1284

Urano K, Kurihara Y, Seki M, Shinozaki K (2010) 'Omics' analyses of regulatory networks in plant abiotic stress responses. Curr Opin Plant Biol 13:132–138

Vandepoele K (2017) A guide to the PLAZA 3.0 plant comparative genomic database. Plant genomics databases. Methods Mol Biol 1533. Humana Press, New York:183–200

Vassilev D, Leunissen JA, Atanassov A, Nenov A, Dimov G (2005) Application of bioinformatics in plant breeding. Wageningen University, Netherland

Vassilev D, Nenov A, Atanassov A, Dimov G, Getov L (2006) Application of bioinformatics infruit plant breeding. J Fruit Ornamental Plant Res 14:145–162

Verma M, Kumar V, Patel RK, Garg R, Jain M (2015) CTDB: an integrated chickpea transcriptome database for functional and applied genomics. PLoS One 10:0136880

Wanchana SS, Thongjuea VJ, Ulat M, Anacleto R, Mauleon M, Conte M, Rouard M, Wang B, Sun YF, Song N, Wei JP, Wang XJ et al (2018) MicroRNAs involving in cold, wounding and salt stresses in Triticumaestivum L., Plant Physiology and Biochemistry In press. Nucleic Acids Res 36:D943–D946

Wang B, Sun YF, Song N, Wei JP, Wang XJ et al (2014) MicroRNAs involving in cold, wounding and salt stresses in *Triticum aestivum* L. Plant Physiol Biochem 80:90–96

Winnenburg R, Baldwin TK, Urban M, Rawlings C, Köhler J, Hammond-Kosack KE (2006) Nucleic Acids Res 34:D459–D464

Wojciech M, Karlowski, Schoof H, Janakiraman V, Stuempflen V, Mayer KFX (2003) MOsDB: an integrated information resource for rice genomics. Nucleic Acids Res 31(1):190–192

Yan S, Du X, Wu F et al (2014) Proteomics insights into the basis of interspecific facilitation formaize (*Zea mays*) in faba bean (Vicia faba)/maize intercropping. J Proteome 109:111–124

Yang X, Li L (2011) miRDeep-P: a computational tool for analyzing the microRNA transcriptome in plants. Bioinformatics 27:2614–2615

Yu J, Hu S et al (2002) A draft sequence of rice genome. Science 296:79–92

Yuan JS, Galbraith DW, Dai SY, Griffin P, Stewart CN Jr (2008) Plant systems biology comes of age. Trends Plant Sci 13:165–171

Zhang Z, Yu J, Li D, Zhang Z, Liu F, Zhou X, Wang T, Ling Y, Su Z (2010) PMRD: plant microRNA database. Nucleic Acids Res 38:D806–D813

Zhang S, Yue Y, Sheng L, Wu Y, Fan G, Li A, Hu X, Shang Guan M, Wei C (2013) PASmiR: a literature-curated database for miRNA molecular regulation in plant response to abiotic stress. BMC Plant Biol 13:1–8

Zhang M, Lv D, Ge P et al (2014) Phosphoproteome analysis reveals new drought response and defense mechanisms of seedling leaves in bread wheat (*Triticum aestivum* L.). J Proteome 109:290–308

Zhou X, Wang G, Sutoh K, Zhu JK, Zhang W (2008) Identification of cold-inducible microRNAs in plants by transcriptome analysis. Biochim Biophys Acta 1779:780–788

Zhou L, Liu Y, Liu Z, Kong D, Duan M et al (2010) Genome-wide identification and analysis of drought-responsive microRNAs in *Oryza sativa*. J Exp Bot 61:4157–4168

Chapter 4
Integration of "Omic" Approaches to Unravel the Heavy Metal Tolerance in Plants

Tanveer Bilal Pirzadah, Bisma Malik, and Khalid Rehman Hakeem

Contents

4.1   Introduction... 79
4.2   Approaches to Study Plant Responses to Stress............................... 81
4.3   How Functional Genomics Play a Role to Combat Abiotic Stress in Plants.............. 81
4.4   Proteomic Tools to Study Heavy Metal Stress in Plants...................... 85
4.5   How Bioinformatics Softwares Play a Lead Role to Study Proteome Maps.............. 86
4.6   Image Analysis Software.. 86
4.7   Mass Spectrometry Analysis Software in Proteomics........................ 88
4.8   Conclusion and Future Perspective.. 89
References.. 90

4.1 Introduction

Abiotic pressure, such as drought, temperature extremes, waterlogging, salinity, heavy metal toxicity, and nutritive deficiency, directly and indirectly alters plant physiology, affecting plant growth and productivity and therefore, causing economic losses in agriculture worldwide. Consequently, the underdeveloped or developing countries discover it extremely difficult to keep countrywide food protection. Climate change might have an effect on all dimensions of food protection, together with food availability, accessibility, usage, and stability. Food production

T. B. Pirzadah (✉)
Department of Bioresources, University of Kashmir, Srinagar, India

Department of Bioresources, Amar Singh College, Cluster University, Srinagar, India

B. Malik
Department of Bioresources, University of Kashmir, Srinagar, India

K. R. Hakeem
Department of Biological Sciences, King Abdulaziz University, Jeddah, Saudi Arabia

Princess Dr. Najla Bint Saud Al-Saud Center for Excellence Research in Biotechnology, King Abdulaziz University, Jeddah, Saudi Arabia

© Springer Nature Switzerland AG 2019   79
K. R. Hakeem et al. (eds.), *Essentials of Bioinformatics, Volume III*,
https://doi.org/10.1007/978-3-030-19318-8_4

agrosystems must be adapted to climate change to ensure food security and stability. Biotechnological and breeding techniques need to reap, maintain, evaluate, file, and disseminate plant genetic assets for various crop species. Adaptive traits with stress-resistant genes need to be diagnosed fast. Breeders must infuse new germplasm into adapted cultivars to enhance productiveness. Discover adaptive traits expressed in exclusive environments more hastily and boom the possibilities of locating genes controlling resistance to heavy metal and different abiotic stress. In an effort to achieve sustainable productivity and improve yield performance, stress-tolerant crops ought to be evolved. Genetic engineering and transgenic techniques have spread out novel ways to diminish the detrimental effect of stress. However, appropriate knowledge of the candidate genes/proteins, their interactions to other genes/proteins, and cross talk to other physiological pathways is needed for developing successful transgenic crops with improved tolerance to stress(s). Techniques combining genomics and proteomics allow for speedy scanning, identification, characterization, and assessment of target genes/proteins for introgression in plants to develop stress tolerance. A number of impressive techniques are available in the functional genomics to decipher functions of genes. At the level of expression, study of all proteins expressed in an organism will be helpful to study the functions of all genes of the organism. Proteomics, which allows for a large-scale examination of proteins, is one of the substantial components of functional genomics. It has gained enormous importance because of its utility in techniques that permits examination of hundreds of proteins in parallel. It is far complementary to genomics, as it targets gene products that are active players of the cell and therefore potential targets for any crop-improvement program in regard to environmental stress (Hossain et al. 2012a). Proteomics has made the approach of reverse genetics truly possible, because by examining the proteins, one can deduce the function of the corresponding gene and the trait that which this gene regulates. Substantial studies have been carried out with impressive achievements in genome and expressed sequence tag (EST) sequencing, containing a wealth of data for many model creatures like plants *Arabidopsis thaliana, Oryza sativa, Zea mays, Lotus japonicus, Hordeum vulgare, Sorghum bicolor, Camellia sinensis,* and *Medicago truncatula.* On the other hand, genome sequence information alone is inadequate to explain the facts concerning gene activity, developmental/regulatory biology, and the biochemical kinetics of life. Proteomic approaches unravel the direct identification and measurement of protein molecules. This specific advantage lets us get over the difficulties associated with disparity between proteomes and genomes, which result from one gene translated into multiple protein products by alternative splicing or posttranslational modifications or expression is spatiotemporally regulated. Therefore, proteome analysis connected to genome sequence information will very likely be highly useful for functional genomics so as to define the function of their associated genes from another aspect (Ohyanagi et al. 2012). In order to better understand the abiotic stress in plants, scientists focus on the multi-omic approaches by combining proteomics with genomics, ionomics, and metabolomics that will give us a clear picture of candidate genes/proteins associated with stress signaling pathways. Proteome variations related to physiological and phenotypic modifications made viable the identity of genes and alleles of interest for development of plants in an effort to keep crop yield as high as possible under adverse environments. Combined approach of omic data is a valuable tool to elucidate the pathways at cellular level (Singh et al. 2016). In the recent past, strategies of protein evaluation accompanied through identity, isolation, cloning, and characterization of genes, promoter analysis, genetic transformation, and new research within the genomic and proteomic platform have culminated in the production of more than a few transgenic for various traits, and they are also correctly applied as evidence-of-concept tools to examine and symbolize the functionality and cross talk of signaling pathways for abiotic stress resistance (Singh et al. 2016). The present review focuses on proteomic profiling of plants in response to HM oxidative stress with emphasis on proteins, especially the antioxidant enzymes and transcription factors, which could be helpful in exploring plant genomic and proteomic elements functional under environmental stress so as to impart tolerance. It additionally analyzes the work carried out thus far, with genomic and proteomic techniques, on plant tolerance to abiotic stress and discusses the various bioinformatics tools for proteome profiling.

4.2 Approaches to Study Plant Responses to Stress

There are two main approaches, specifically, functional genomics and proteomics, to observe the reaction of plants to abiotic stress conditions (Fig. 4.1). Functional genomic targets the evaluation of transcripts with the intention to analyze differentially transcribed mRNAs within stressed plants. The technique is called as DNA microarrays in which RNAs are used to synthesize cDNAs for further analysis. On the other hand, proteomics evaluates proteins with the purpose to target and recognize differentially translated (synthesized) necessary protein in stressed plants over the control ones, and the technique is referred as protein microarrays. The functional genomics and proteomics thus constitute collectively an important tool reverse genetics (Fig. 4.2).

4.3 How Functional Genomics Play a Role to Combat Abiotic Stress in Plants

Molecular biology offers several strategies to assess gene characteristics, and efforts had been made to broaden novel strategies to illustrate the expression and characteristic of unique genes. Functional genomics is right now widely seen as offering satisfactory tools for looking into the abiotic stress reactions in plants through which usually networks of stress perception, signal transduction, and protective responses can be analyzed from gene transcription, via protein complements of cells, to the metabolite information of stressed tissues. Functional genomics employs multiple parallel approaches, including global transcript profiling coupled with using mutants and transgenics, in order to study gene function within a high-throughput setting. The provision of a huge volume of genomic data (database) has provided

Fig. 4.1 Integrated omic approach to study heavy metal stress in plants. Genomics, proteomics, and bioinformatics are useful tools that can help us to elucidate and analyze active regulatory networks controlling heavy metal stress responses and tolerance

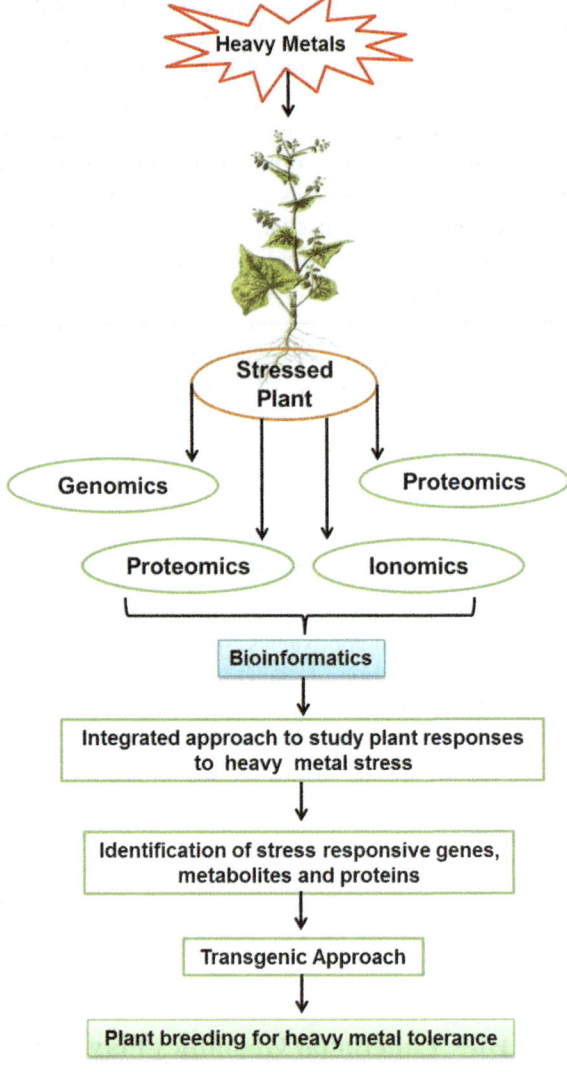

information regarding the gene content of plants. Partial or whole sequences regarding cDNA often provide a firm basis of the dimension of the transcriptome, typically the total mRNA content in the organism at given moment. Three important databases, viz. National Center for Biotechnology Information (NCBI) UniGene, http://www. ncbi.nlm.nih.gov/; The Institute of Genomic Research (TIGR) Gene Indices, http:// www.tigr.org; and Sputnik, http://mips.gsf.de/proj/sputnik, serve to manage the available plant expression sequence tags (ESTs), collectively with properly charac- terized genes, into nonreductant gene clusters. ESTs have formed the key of studies within the international gene expression of numerous stress tolerance traits in some plants including *Arabidopsis* and rice (Rensink et al. 2005). The evaluation of ESTs

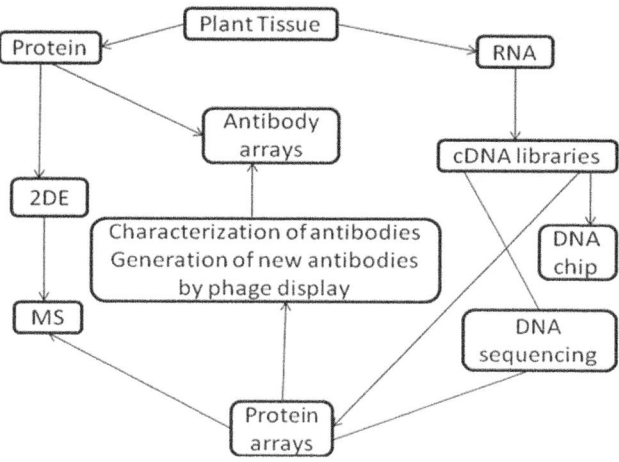

Fig. 4.2 Genomics and their correlation to proteomics

produced from cDNA libraries of salt-stressed rice showed that there was an increment in transcripts associated with cell rescue, protection, transport, energy, and metabolism, but the majority of stress-inducible genes could not be assigned any activity (Bohnert et al. 2001). In a study performed to identify salt-stress-inducible ESTs obtained from polymerase chain reaction (PCR) subtraction in salt-tolerant rice, 384 genes were recognized as salt responsive, <5% of which have been additionally confirmed by Northern blotting. Almost 50% of these genes were recognized for involvement in detoxification, stress response, growth, and development (Shiozaki et al. 2005). Baisakh et al. (2008) used cDNA-AFLP as an RNA imaging technique to discover transcripts that accumulated strongly and have been triggered de novo in *Spartina alterniflora* (smooth cordgrass) in response to salinity stress. Of the 213 cDNAs isolated, 28 have been diagnosed belonging to diverse sets of genes involved inside ion transport and compartmentalization, cell division, and metabolism, in addition to protein synthesis. At distinct degrees of salt stress, the expression patterns of 14 such genes were analyzed via RT-PCR, and their direct and oblique relationship with salinity tolerance was established. Apart from cDNA-AFLP, differential display approach is based on typically the synthesis of cDNA using an oligo dT primer (3′) and an arbitrary oligonucleotide primer for the 5′-end and amplifies rare cDNA. In *Gossypium arboreum*, Venkatesh et al. (2005) used the differential display reverse transcriptase (DDRT) PCR to compare the total dissimilarities in gene expression between water-stressed and control plants. By screening 93 primer-pair combinations, the DDRT approach brought out upregulation of 30 cDNA transcripts. By means of reamplification and quality control assay, 10 cDNA transcripts appeared false positive, while the rest of the 20 cDNA transcripts had been excised from the gel, reamplified, cloned, and sequenced. Homology search discovered that six transcripts confirmed extensive homology with known genes. RT PCR confirmed that, among six transcripts, five had been substantially overexpressed in water-stressed leaves with respect to control. This finding is crucial, due to the fact there is only couple of reports of the common stress protein; transposable factors were available in plants but none in cotton under drought stress. Serial analysis of gene expression (SAGE) is likewise a dominant technique that can be used to quantify the global gene expression. The tool, developed first to quantify gene expression in yeast, includes production of brief 10–14 nucleotide tags, with every tag representing a unique transcript present within a cell (Velculescu et al. 1995). Determining the sequence of a tag concedes for recognition of the analogous gene, and the frequency regarding tag represents the stable state of the mRNA from which it is extracted. SAGE technique was utilized for the quantification of global gene expression especially in rice by Matsumura et al. (1999); 10,122 tags from 5921 expressed genes of rice seedlings were analyzed, 18 genes had been determined to be anaerobically induced, and 6 genes were suppressed. The anaerobically induced genes had been those coding for prolamin, expansins, and glycine-rich cellular wall protein. Jung et al. (2003) carried out an experiment using SAGE technique to unravel the changes in gene expression in the leaves of *Arabidopsis* subjected to cold stress. For the transcription profiling on a genome-wide scale, tools like massive parallel signature sequencing (MPSS) are also used. This tool, like SAGE, can be used to acquire an illustration of the mRNA population in a sample, which can be associated with ESTs, mRNA, or the whole genome sequence, but the magnitude of the generated data is much larger (Brenner et al. 2000). The MPSS resource for rice consists of 20 MPSS libraries generated from varied tissues, including 3 from abiotic stress, particularly cold, dehydration, and salt (Nakano et al. 2006). Recently, a DNA-chip technological innovation, based on the ability to bind either DNA fragments or previously characterized oligonucleotides on a microscope slide, has been developed to examine gene expression profiles. To examine and compare the global gene expression patterns especially in plants like rice, maize, strawberry, petunia, ice plants, and lime bean, cDNA and oligonucleotide microarrays have been widely used. Kawasaki et al. (2001) first reported the use of microarray technique to study the global gene expression among two rice varieties (var. Pokkali-salt-tolerant and var. IR29-salt sensitive) subjected to salt stress. The evaluation was carried out using cDNA microarray comprising 1728 cDNA clones developed from control or salt-stressed roots of Pokkali. The principle distinction between the expression patterns of the two varieties was the deferred timing of the IR29 response in terms of kinetics of gene expression, which could be responsible for its salt sensitivity (Kawasaki et al. 2001). An additional approach for functional genomics is the utilization of insertional mutagenesis, which involves possibly transposon mutagenesis or T-DNA mutagenesis. In both instances, a known DNA sequence is introduced, which might be randomly inserted at many places within genome that in turn causes loss of function mutation or gene knockouts and rarely develop gain-of-function mutation. Insertional mutagenesis has been employed broadly to signify the abiotic stress-responsive genes, such as those coding for AtHKT1 (a high-affinity potassium transporter), CIPK3 (calcium-associated protein kinase), CBL1 (calcineurin B-like protein kinase), OSM1/SYP61 (syntaxin), and HOS10 (R2R3-type MYB transcription factor) (Rus et al. 2001; Zhu et al. 2002, 2005; Cheong et al. 2003; Kim et al. 2003). As a result the strategies associated with the functional genomics are effective and high yielding molecular-biology techniques, which are currently being used to identify the transcripts susceptible to the abiotic stress.

4.4 Proteomic Tools to Study Heavy Metal Stress in Plants

High-throughput omic tools are thoroughly being exploited these days to dissect plants' molecular strategies of heavy metal (HM) stress tolerance. The plants growing in heavy metal contaminated sites have the innate ability to develop homeostatic mechanisms to modify the uptake, mobilization, and intracellular toxic metal concentration to alleviate oxidative stress. As the functional translated portion of the genome plays an essential role in plant stress response, proteomic research caters a finer picture of protein networks and metabolic pathways mostly involved in cellular detoxification and tolerance mechanism against HM toxicity. Over the past decade, thorough research on plants' response to HM stress has been conducted to unravel the tolerance mechanism. Genomic technologies have been useful in addressing abiotic stress responses in plants, including HM toxicity (Bohnert et al. 2006). One of the drawbacks of genomic approach is that the changes in gene expression at the transcript level have not always been reflected at protein level (Gygi et al. 1999). Thus, an exhaustive proteomic evaluation is of great significance to perceive target proteins that are actively involved in HM detoxification. It is reported that plant response to HM stress has been reviewed extensively during the last decade (Hossain et al. 2012b). Proteomic approach helps to identify the proteins in plants subjected to HM stress and their role in mitigation. Majority of the proteomic studies carried out thus far on HM-associated toxicity disclose positive correlation among tolerance and increased abundance of scavenger proteins, for instance, increased expression of superoxide dismutase (SOD) isoforms (Cu/Zn-SOD, Fe-SOD) in plants subjected to Cd (Hossain et al. 2012b) and Al stress (Yang et al. 2007). Alvarez et al. (2009) carried out a proteome analysis of *B. juncea* exposed to Cd stress and found upregulation of only Fe-SOD, while as Cu/Zn-SOD gets downregulated and at the same time, this protein is considered to play a pivotal role in the defense system. An exhaustive proteomic work conducted by Hajduch et al. (2001) in rice seedlings subjected to HM stress unravels the drastic reduction in abundance/fragmentation of a number of subunits of RuBisCO small subunit (SSU) and large subunit (LSU), suggesting thorough interruption of photosynthetic machinery. Proteomic study by Lee et al. (2010) discovered induction of vacuolar proton-ATPase in rice roots and leaves depicting their positive role in Cd detoxification through vascularization. Shin et al. (2010) performed the proteomic evaluation of two buckwheat species under light and dark conditions and concluded that the sprouting leaves of 7-day-old etiolated common buckwheat seedlings turned to light yellow, suggesting the inhibition of light-dependent protochlorophyllide reductase by proteome evaluation. Duressa et al. (2011) carried out a proteomic evaluation among Al-tolerant and Al-sensitive soybean genotypes and concluded that there occurs accumulation of enzymes in the Al-tolerant genotypes that are involved in the synthesis of citrate-a the main player in the detoxification of Al metal ions, whereas in case of Al-sensitive genotypes, general stress proteins get induced. Bagheri et al. (2015) carried out a comparative proteomic evaluation in spinach to better understand the protein players responding not only to particular stress of Cd or salinity but also to their combination. They reported that proteome modulation, by all combinations (metal and salinity) of stressors, signifies that spinach does not depend on any particular set of proteins and pathways, but a multitude of responses is initiated. Upon enhancing tolerance to stress, the plant can induce novel proteins and sometimes inhibit the synthesis of others after equipping itself in an appropriate way against any specific or a group of stresses. Because the information on functioning mechanisms of most stress proteins provides planning strategies for reconstructing stress tolerance via transgenic approach, it is the need of the hour to identify, isolate, and characterize a number of stress proteins. The stress-responsive proteins act as important elements of abiotic stress tolerance, and a further study related to their functions would lead to a deeper insight with novel discoveries related to abiotic stress tolerance in plants.

4.5 How Bioinformatics Softwares Play a Lead Role to Study Proteome Maps

To unravel the proteome profile of the sample, the primary step involves fractionation of the sample by 2-D gel activity. The distinct protein spots are removed, digested by trypsin, and then analyzed by mass spectrometry followed by database searching for protein identification. In major proteomic projects, the data on samples used in 2-D gel is usually recorded as a part of a laboratory information management system (LIMS). The other access to 2-D gel analysis sometimes involves direct digestion of the protein or protein mixture with a site-specific protease, viz., trypsin, followed consecutively by liquid chromatography (LC) or capillary electrophoresis (CE) conjugated with online tandem mass spectrometry (MS/MS).

4.6 Image Analysis Software

Image analysis software is a crucial tool to generate images of the sample gels and was introduced in the mid-1970s, and additional advancements were created within the late 1980s (Henzel et al. 1993). Few commercially accessible 2-D gel image analysis softwares are presented in Table 4.1. These softwares are devised with the aim to digitize images from 2-D gels, identify and quantify spots of different intensities among gels by matching the control and the sample gel(s), and develop reports listing the differentially expressed protein spots. However, the software depends upon various parameters, viz., the computer program, the file format, or

Table 4.1 Some commercially available 2-D gel image analysis software

Name of software	Source
DeCyder™ 2D Analysis	Amersham Biosciences (www.apbiotech.com)
Delta2D™ (www.decodon.com/Solutions/Deltta2D.html)	DECODON GmbH (www.decodon.com)
GELLAB II+ (www.scanalytics.com/product/gel/Gellab.html)	Scanalytics (www.scanalytics.com)
GD Impressionist™ (www.genedata.com/products/impressionist)	GeneData (www.genedata.com)
ImageMaster™ 2D Elite	Amersham Biosciences (www.apbiotech.com)
ImagepIQ™ (www.proteomesystems.com/product/profile.asp?Category=gel+image+analysis)	Proteome Systems Ltd. (www.proteomesystems.com)
Investigator™ HT PC Analyzer[a] (www.genomicsolutions.com/proteomics/2dgelanal.html)	Genomic Solutions (www.genomicsolutions.com)
Melanie 3 (www.genebio.com/Melanie.html)	GeneBio (www.genebio.com)
PDQuest™ 2-D Analysis Software	Bio-Rad Laboratories (www.bio-rad.com)
Phoretix™ 2D (www.phoretix.com/products/2d_products.htm)	Nonlinear Dynamics (www.nonlinear.com)
Progenesis™ (www.nonlinear.com/2D/progenesis)	Nonlinear Dynamics (www.nonlinear.com)
ProteinMine™ (www.scimagix.com/products-ProteinMine.html	Scimagix (www.scimagix.com)
ProteomeWeaver (www.definiens-imaging.com/proteomeweaver)	Definiens Imaging GmbH (www.defeniens-imaging.com)
TotalLab (www.totallab.com)	Nonlinear Dynamics (www.nonlinear.com)
Z3 2D-Gel Analysis System (www.2dgels.com)	Compugen Ltd. (www.compugen.co.il)
ImageQuant TL	GE Healthcare
Quantity One	Bio-Rad Laboratories (www.bio-rad.com)
GelAnalyzer	Freely available on WWW
GelScape	Freely available on WWW
ImageMaster 2D Platinum	GE Healthcare
REDFIN	Ludesi
Dymension	Syngene
Proteovue	Eprogen

[a]Developed in collaboration with nonlinear dynamics

the graphical user interface. Another step in proteomics is the protein identification by means of mass spectrometry. Few 2-D gel image analysis packages are accompanied with automated robotic systems to direct the excision of spots of choice for subsequent MS analysis. For putting in place an advert or high-throughput technique for proteomics, it's mandatory that the softwares for 2-D gel image analysis support robust database tools for storing images and identification and quantification of data from a particular protein spot. However, several 2-D gel image analysis softwares lack sturdy integration with the powerful relative databases that presently store most enterprise data. The different proprietary storage formats employed at commercial level further complicate the accessibility and online handiness of helpful data generated by the image analysis software. In the present scenario, for instance, Scimagix (San Mateo, CA, USA) has developed its 2-D gel image analysis product, ProteinMine®, on the commercial Oracle® relative database system (Oracle Corporation, Redwood Shores, CA, USA) used by the company for their scientific image management system (SIMS). This sort of open storage design may be of significant value to laboratories seeking to integrate with other LIMS packages they are using.

4.7   Mass Spectrometry Analysis Software in Proteomics

The MS analysis wasn't appropriate for the biomolecules till the invention of soft ionization techniques like MALDI (matrix-assisted laser desorption ionization) (Tanaka et al. 1988) and ESI (electrospray ionization) (Yamashita and Fenn 1984). While MALDI is appropriate for solid samples, ESI is appropriate for the samples dissolved in distinct solvents. MALDI or ESI sources could be conjugated with various types of analyzers like ToF (time of flight), LTQ (linear trap quadrupole), Orbitrap, and Quadrupole to attain distinct types of MS analysis with varying sensitivity and precision. The mass spectrum of every protein species is unique, and because of such uniqueness, it is known as peptide mass fingerprint (PMF) (Pappin et al. 1993). PMFs are searched in a database by using software like Mascot® or Sequest® to find out the identity of a particular protein. The PMF only includes masses of the peptides only but lacks amino acid sequences. The success rate of the PMF matches to the database can be enhanced by the addition of amino acid sequence data obtained by collision-induced dissociation (CID) mechanism by which the gas-phase ions are further fragmented by colliding neutral molecules, viz., helium, argon, and nitrogen. With the help of CID from the fragments of the peptides, amino acid sequences could be predicted. The PMF data in conjugation with sequence information results in higher protein identification. To obtain a CID data, an MS system has to have two mass analyzer divided by a collision-induction cell, and this type of MS systems is known as MS/MS or tandem MS (Wells and McLuckey 2005). Table 4.2 represents the list of several software tools for database searches using MS data for protein identification. Most of these softwares have free access over the Internet, but we are able to conjointly use them offline through a sound license. Among the tools depicted in Table 4.2, Mascot and Sequest can be run on computer clusters. Even though the details regarding the computer clusters and computer farms in large-scale industry-based proteomics are often proprietary, some data is available on their scope and magnitude. For example, it has been reported that GeneProt's large-scale proteomic discovery center in Geneva,

Table 4.2 Some database search tools for protein identification based on mass spectrometric data

Name of software	Source
Peptide mass fingerprint or peptide mass map analysis	
Mascot[a] (www.matrixscience.com)	Matrix Science Ltd.
MassSearch[a] (cbrg.inf.ethz.ch/Server/MassSearch.html)	ETH
Mowse[a] (www.hgmp.mrc.ac.uk/Bioinformatics/Webapp/ mowse)	UK Human Genome Mapping Project Resource Centre
MS-FIT[a] (prospector.ucsf.edu)	UCSF
*Pep*MAPPERa (wolf.bms.umist.ac.uk/mapper)	UMIST
PepSea[a] (195.41.108.38/PepSeaIntro.html)	Protana
PeptideSearch[a] (www.narrador.embl-heidelberg.de/ GroupPages/PageLink/peptidesearchpage.html)	EMBL
PeptIdent[a,b] (www.expasy.ch/tools/peptident.html)	ExPASy
ProFound[a] (prowl.rockefeller.edu/PROWL/prowl.html), (also at service.proteometrics.com/prowl/profound.html)	Rockefeller University and Proteometrics
Peptide sequence or peptide sequence tag query	
Mascot[a] (www.matrixscience.com)	Matrix Science Ltd.
MS-Seq[a] (prospector.ucsf.edu)	UCSF
PepSea[a] (195.41.108.38/PepSeaIntro.html)	Protana
PeptideSearch[a] (www.narrador.embl-heidelberg.de/. GroupPages/PageLink/peptidesearchpage.html)	EMBL
TagIdent[a] (www.expasy.ch/tools/tagident.html)	ExPASy
MS/MS ion search analysis	
Mascot[a] (www.matrixscience.com)	Matrix Science Ltd.
MS-Tag[a] (prospector.ucsf.edu)	UCSF
PepFrag[a] (prowl.rockefeller.edu/PROWL/prowl.html)	Rockefeller University
Sequest[c] (fields.scripps.edu/sequest)	Scripps Research Institute
Sonars MS/MS[a] (www.proteometrics.com)	Proteometrics

[a]Available for data search over the Internet
[b]PeptIdent2 (now called SmartIdent) is a completely different tool from PeptIdent. PeptIdent2 utilizes a rather unique approach involving a generic algorithm using a training set of protein mass spectra. However, it is not yet available on the publicly available ExPASy server
[c]Sequest is commercially available through Thermo Finnigan, San Jose, CA, USA

Switzerland, will use 1420 Compaq® Alpha-based Tru64 UNIX computer processors to capture, store, and analyze the terabytes of proteomic data that will be generated by 51 mass spectrometers (Genome Web News 2001).

4.8 Conclusion and Future Perspective

About 3.1 billion individuals from developing countries reside in rural areas, and out of this population, ~2.5 billion individuals rely on agrarian practices for their livelihood, which constitutes 30% of economic growth because of the gross domestic product obtained from agricultural sector (FAO 2012). World population is gaining a rapid momentum, and it is expected that by the middle of the twenty-first century, it will be around 10 billion which in turn leads to severe food crises all over the world. Moreover, due to the increasing trend in various anthropogenic activities, the quality of the environment gets deteriorated that pose a great threat to the global crop productivity, and therefore, it is the need of the hour to select such crops that are going to adapt and resist enormous abiotic stress conditions. Abiotic stress tolerance in plants is not regulated by a monogenic trait but is complex and multigenic in nature and involves different signaling components and therefore is more challenging to control and engineer. Thus, engineering approaches for heavy metal resistance involve expression of genes, whose metabolite(s) play a pivotal role in signaling and regulate the biosynthesis of various compounds that impart the heavy metal tolerance in plants. Nowadays research should be focused on signal transduction pathways induced by HMs as they can utilize common signal elements that can also be elicited by other environmental stresses to better understand the metal homeostasis. In the future, multiple stress factors will be investigated as it happens in real environmental conditions. An interdisciplinary approach with well-integrated "omics," viz., genomics, proteomics, ionomics, metabolomics, and bioinformatics, is necessary to unravel the molecular mechanisms involved in HM stress; besides, it provides a concrete understanding of the metallo-proteome synergistic action.

References

Alvarez S, Berla BM, Sheffield J, Cahoon RE, Jez JM, Hicks LM (2009) Comprehensive analysis of the *Brassica juncea* root proteome in response to cadmium exposure by complementary proteomic approaches. Proteomics 9(9):2419–2431

Bagheri R, Bashir H, Ahmad J, Iqbal M, Qureshi MI (2015) Spinach (*Spinacia oleracea* L.) modulates its proteome differentially in response to salinity, cadmium and their combination stress. Plant Physiol Biochem 97:235–245

Baisakh N, Subudhi PK, Varadwaj P (2008) Primary responses to salt stress in a halophyte, smooth cordgrass (*Spartina alterniflora* Loisel.). Funct Integr Genomics 8:287–300

Bohnert HJ, Ayoubi P, Borchert C, Bressan RA, Burnap RL, Cushman JC, Cushman MA, Deyholos M, Fischer R, Galbraith DW et al (2001) A genomics approach towards salt stress tolerance. Plant Physiol Biochem 39:295–311

Bohnert HJ, Gong Q, Li P, Ma S (2006) Unraveling abiotic stress tolerance mechanisms–getting genomics going. Curr Opin Plant Biol 9:180–188

Brenner S, Johnson M, Bridgham J, Golda G, Lloyd DH, Johnson D, Luo S, McCurdy S, Foy M, Ewan M et al (2000) Gene expression analysis by massively parallel signature sequencing (MPSS) on microbead arrays. Nat Biotechnol 18:630–634

Cheong YH, Kim KN, Pande GK, Gupta R, Grant JJ, Luan S (2003) CBL1, a calcium sensor that differentially regulates salt, drought, and cold responses in Arabidopsis. Plant Cell 15:1833–1845

Duressa D, Soliman K, Taylor R, Senwo Z (2011) Proteomic analysis of soybean roots under aluminum stress international. Int J Plant Genomics 2011:2825–2831

FAO (2012) Statistical yearbook Viale delle Terme di Caracalla. Rome ISBN 978-92-5-107083-3

GenomeWeb News (2001) Backed by Compaq, Bruker, and Novartis, Gene-Prot opens industrial-scale proteomics facility. 26 April (www.genomeweb.com)

Gygi SP, Rist B, Gerber S, Turecek F, Gelb MH, Aebersold R (1999) Quantitative analysis of complex protein mixtures using isotope-coded affinity tags. Nat Biotechnol 17:994–999

Hajduch M, Rakwal R, Agrawal GK, Yonekura M, Pretova A (2001) High-resolution two-dimensional electrophoresis separation of proteins from metal-stressed rice (Oryza sativa L.) leaves: drastic reductions/fragmentation of ribulose-1,5-bisphosphatecarboxylase/oxygenase and induction of stress-related proteins. Electrophoresis 22:2824–2831

Henzel WJ, Billeci TM, Stults JT, Wong SC, Grimley C, Watanabe C (1993) Identifying proteins from two-dimensional gels by molecular mass searching of peptide fragments in protein sequence databases. Proc Natl Acad Sci U S A 90:5011–5015

Hossain MA, Piyatida P, Teixeirada Silva JA, Fujita M (2012a) Molecular mechanism of heavy metal toxicity and tolerance in plants: central role of glutathione in detoxification of reactive oxygen species and methylglyoxal and in heavy metal chelation. J Bot:1–37

Hossain Z, Hajika M, Komatsu S (2012b) Comparative proteome analysis of high and low cadmium accumulating soybeans under cadmium stress. Amino Acids 43:2393–2416

Jung SH, Lee JY, Lee DH (2003) Use of SAGE technology to reveal changes in gene expression in Arabidopsis leaves undergoing cold stress. Plant Mol Biol 52:553–567

Kawasaki S, Borchert C, Deyholos M, Wang H, Brazille S, Kawai K, Galbraith D, Bohnert HJ (2001) Gene expression profiles during the initial phase of salt stress in rice. Plant Cell 13:889–905

Kim KN, Cheong YH, Grant JJ, Pandey GK, Luan S (2003) CIPK3, a calcium sensor-associated protein kinase that regulates abscisic acid and cold signal transduction in Arabidopsis. Plant Cell 15:411–423

Lee K, Bae DW, Kim SH, Han HJ, Liu X, Park HC et al (2010) Comparative proteomic analysis of the short-term responses of rice roots and leaves to cadmium. J Plant Physiol 167:161–168

Matsumura H, Nirasawa S, Terauchi R (1999) Transcript profiling in rice (Oryza sativa L.) seedlings using serial analysis of gene expression (SAGE). Plant J 20:719–726

Nakano T, Suzuki K, Fujimura T, Shinshi H (2006) Genome-wide analysis of the ERF gene family in Arabidopsis and rice. Plant Physiol 140:411–432

Ohyanagi H, Sakata K, Komatsu S (2012) Soybean proteome database 2012: update on the comprehensive data repository for soybean proteomics. Front Plant Sci 3:110

Pappin DJ, Hojrup P, Bleasby AJ (1993) Rapid identification of proteins by peptide-mass fingerprinting. Curr Biol 3(6):327–332

Rensink WA, Lee Y, Liu J, Iobst S, Ouyang S, Buell CR (2005) Comparative analyses of six solanaceous transcriptomes reveal a high degree of sequence conservation and sequence-specific transcripts. BMC Genomics 6:124

Rus A, Yokoi S, Sharkhuu A, Reddy M, Lee BH, Matsumoto TK, Koiwa H, Zhu JK, Bressan RA, Hasegawa PM (2001) AtHKT1 is a salt tolerance determinant that controls Na+ entry into plant roots. Proc Natl Acad Sci U S A 98:14150–14155

Shin DH, Kamal AHM, Suzuki T, Yun YH, Lee MS, Chung KY, Jeong HS, Park CH, Choi JS, Woo SH (2010) Reference proteome map of buckwheat (Fagopyrum esculentum and Fagopyrum tataricum) leaf and stem cultured under light or dark. Aust J Crop Sci 4(8):633–641

Shiozaki N, Yamada M, Yoshiba Y (2005) Analysis of salt-stress inducible ESTs isolated by PCR-subtraction in salt-tolerant rice. Theoretical Appl Genet 110:1177–1186

Singh S, Parihar P, Singh R, Singh VP, Prasad SM (2016) Heavy metal tolerance in plants: role of transcriptomics, proteomics, metabolomics, and ionomics. Front Plant Sci 6:1143

Tanaka K, Waki H, Ido Y, Akita S, Yoshida Y, Yoshida T (1988) Protein and polymer analyses up to m/z 100 000 by laser ionization time-of-flight mass spectrometry. Rapid Commun Mass Spectrom 2:151–153

Velculescu VE, Zhang L, Vogelstein B, Kinzler KW (1995) Serial analysis of gene expression. Science 270(5235):484–487

Venkatesh B, Hettwer U, Koopmann B, Karlovsky P (2005) Conversion of cDNA differential display results (DDRT-PCR) into quantitative transcription profiles. BMC Genomics 6:51

Wells JM, McLuckey SA (2005) Collision-induced dissociation (CID) of peptides and proteins. Methods Enzymol 402:148–185

Yamashita M, Fenn JB (1984) Electrospray ion source. Another variation on the free-jet theme. J Phys Chem 88(20):4451–4459

Yang Q, Wang Y, Zhang J, Shi W, Qian C, Peng X (2007) Identification of aluminum-responsive proteins in rice roots by a proteomic approach: cysteine synthase as a key player in Al response. Proteomics 7:737–749

Zhu J, Gong Z, Zhang C, Song CP, Damsz B, Inan G, Koiwa H, Zhu JK, Hasegawa PM, Bressan RA (2002) OSM1/SYP61 a syntaxin protein in arabidopsis controls abscisic acid–mediated and non-abscisic acid–mediated responses to abiotic stress. Plant Cell 14(12):3009–3028

Zhu J, Verslues PE, Zheng X, Lee BH, Zhan X, Manabe Y, Sokolchik I, Zhu Y, Dong CH, Zhu JK, Hasegawa PM, Bressan RA (2005) Proc Natl Acad Sci U S A 102(28):9966–9971

Chapter 5
Advanced Multivariate and Computational Approaches in Agricultural Studies

Inayat Ur Rahman ⓘ**, Eduardo Soares Calixto, Aftab Afzal, Zafar Iqbal, Niaz Ali, Farhana Ijaz, Muzammil Shah, and Khalid Rehman Hakeem**

Contents

5.1	Introduction	93
5.2	Methodology	94
5.3	Ordination Analyzes	95
5.4	Correlograms, Heatmaps, and Scatterplot Matrix	96
5.5	Violin and Box Plot	96
5.6	Chord Diagram and Bipartite Networks	99
5.7	Hierarchical Clustering	101
5.8	Final Remarks	101
References		102

5.1 Introduction

The statistical principles underlying design of experiments were pioneered by R. A. Fisher in the 1920s and 1930s at Rothamsted Experimental Station, an agricultural research station around 40 km north of London. Fisher had shown the way on how

I. U. Rahman (✉)
Department of Botany, Hazara University, Mansehra, Khyber Pakhtunkhwa, Pakistan

William L. Brown Center, Missouri Botanical Garden, St. Louis, MO, USA

E. S. Calixto
Department of Biology, University of São Paulo, São Paulo, Brazil

University of Missouri St. Louis (UMSL), Saint Louis, MO, USA

A. Afzal · Z. Iqbal · N. Ali · F. Ijaz
Department of Botany, Hazara University, Mansehra, Khyber Pakhtunkhwa, Pakistan

M. Shah
Department of Biological Sciences, King Abdulaziz University, Jeddah, Saudi Arabia

K. R. Hakeem
Department of Biological Sciences, King Abdulaziz University, Jeddah, Saudi Arabia

Princess Dr. Najla Bint Saud Al-Saud Center for Excellence Research in Biotechnology, King Abdulaziz University, Jeddah, Saudi Arabia

© Springer Nature Switzerland AG 2019
K. R. Hakeem et al. (eds.), *Essentials of Bioinformatics, Volume III*,
https://doi.org/10.1007/978-3-030-19318-8_5

to draw valid conclusions from field experiments where nuisance variables such as temperature, soil conditions, and rainfall are present. He had shown that the known nuisance variables usually cause systematic biases in results of experiments and the unknown nuisance variables usually cause random variability in the results and are called inherent variability or noise. He introduced the concept of analysis of variance (ANOVA) for partitioning the variation present in data due to (a) attributable factors and (b) chance factors. The methodologies he and his colleague Frank Yates developed are now widely used. No doubt, these methodologies have a profound impact on agricultural sciences research.

It may be emphasized in the beginning itself that experimental design is first about agriculture, animal science, biology, chemistry, industry, education, etc. and then about statistics and mathematics. In fact, experimental design forms the backbone of agricultural sciences; it is an integral component of every research endeavor in agricultural sciences. To design a good experiment, the researcher first needs to outline questions to be answered or needs one or more well-defined hypotheses.

Therefore, the application of advanced multivariate analyses and decision support tools or approaches are necessary and suggested so that researchers can better evaluate and present the results of different studies related to agricultural system. Thus the aim of this study is to provide a new support for analyses based on advanced multivariate approaches, concerning to the main statistical analyses used in agricultural system, as well as to provide better ways to show and manipulate the data graphically, enhancing the publication potential of the studies.

5.2 Methodology

Data used in this chapter were acquired from agricultural studies with the authors' authorization. For Sects. 5.4 and 5.5, we used data related to the chickpea (*Cicer arietinum* Medik) production. In this study, the authors analyzed the chickpea growth based on different soil treatments. The five treatments were (A) diammonium phosphate (DAP) half dose (12 g) + biofertilizer; (B) ammonium molybdate (0.236 g) + zinc sulphate (0.096 g) + biofertilizer; (C) ammonium molybdate (0.165 g) + zinc sulphate (0.144 g) + biofertilizer; (D) ammonium molybdate (0.236 g) + zinc sulphate (0.096 g); and (E) ammonium molybdate (0.165 g) + zinc sulphate (0.144 g). The parameters analyzed were plant height (PH) (cm), root length (RL) (cm), plant fresh weight (PFW) (g), plant dry weight (PDW) (g), root fresh weight (RFW) (g), number of flowers per plant (NoFPP), number of pods per plant (NoPPP), number of branches per plant (NoBPP), stem diameter (SD) (cm), and number of leaves per plant (NoLPP). For Sect. 5.6, we used data-related bands and genotypes.

All analyses and graphs elaborated were assembled in CANONO and RStudio 3.5.1 software at a level of 5% significance. The packages and main functions used in R are described in each topic addressed.

5.3   Ordination Analyzes

Overview Ordination techniques are used to describe relationships between dependent and independent variables. Principal Components Analysis (PCA) is one of the earliest ordination technique invented by Karl Pearson in 1901 (Dunn and Stearns 1987). Currently, it is mostly used as a tool in exploratory data analysis and for making predictive models. PCA uses a rigid rotation to derive orthogonal axes, which maximize the variance in the data set. Computationally, Principal components analysis is the basic eigen analysis technique. It maximizes the variance explained by each successive axis. The sum of the eigen values will equal the sum of the variance of all variables in the data set. PCA is relatively objective and provides a reasonable but crude indication of relationships i.e. in an indirect non-canonical way.

Example From PCA (Fig. 5.1), three groups, A–B, C–D, and E can be distinguished. These groups are separated by the analyzed characteristics of each treatment, where NoLPP (number of leaves per plant) and PH (plant height) seem to be the main factors that explain this grouping.

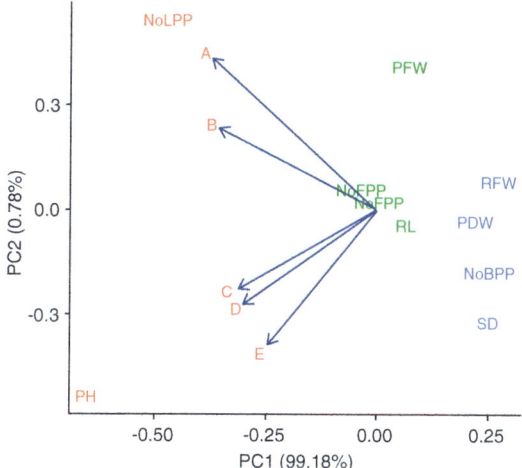

Fig. 5.1 PCA showing correlation of treatments in three groups (A–B, C–D, and E) based on the characteristics analyzed for each treatment, where NoLPP and PH seem to have a major influence in this grouping. *PH* plant height (cm), *RL* root length (cm), *PFW* plant fresh weight (g), *PDW* plant dry weight (g), *RFW* root fresh weight (g), *NoFPP* number of flowers per plant, *NoPPP* number of pods per plant, *NoBPP* number of branches per plant, *SD* stem diameter (cm), *NoLPP* number of leaves per plant. *Packages*: ggfortify (Tang et al. 2016), cluster (Maechler et al. 2018); *Functions*: autoplot, clara

5.4   Correlograms, Heatmaps, and Scatterplot Matrix

Overview Correlogram is a kind of correlation matrix, which shows the relationship between each pair of numerical variables analyzed based on the degree of association. Heatmap (Fig. 5.2) is a graphical tool based on a color-coding system to represent the relationship between pairs of variables and calculated by dissimilarity. A scatterplot matrix (Fig. 5.2) is a set of scatterplots organized in a matrix or grid and shows the relationship between pairs of variables. Scatterplot matrix is very useful for exploratory data analysis, especially for linear correlation between multiple variables.

Example From correlogram (Fig. 5.2, top left), it may be observed that in general there is a strong correlation among all parameters analyzed, except for SD. The lower part of the correlogram is the R^2 values, the diagonal is the parameter names, and the upper part is pie graph showing the same relation presented in the lower part. The color in both lower and upper parts represents the positive and negative relationship. In this example, we can see that blue color shows a positive relation, red color presents a negative relation, and white color represents no relation. The heatmap shows the relation among the parameters through distance of dissimilarity (in our case, Euclidian distance). Three different clusters may be seen in the heatmap of Fig. 5.2, top right. The first cluster, in green, is represented by treatments A and B (see Methodology for differences among treatment), the second is represented by treatments C and D, and the third is represented by treatment E. Thus, from the ten parameters analyzed, differences among treatments may be seen, and the chickpea growth has been strongly influenced by soil components (see Sect. 5.5). Lastly, similar to correlogram, the scatterplot (Fig. 5.2, bottom) shows the relation between two variables, where we can see the absolute value (correlation coefficient – "r") of the correlation between variables and the significance of the relationship discriminated by asterisks ($***p < 0.001$, $**p < 0.01$, $*p < 0.05$) on top, the bivariate scatterplots with fitted line on bottom, and histograms with kernel density estimation and rug plot on diagonal (Fig. 5.2, bottom). For instance, we can observe that SD presents a low correlation value with all other parameters.

5.5   Violin and Box Plot

Overview Box plot is a simple and standardized way to display data distribution. This plot is usually based on five elements: minimum, maximum, first and third quartile, and median. In addition, individual points can be plotted, especially if the data present outliers. This kind of graphic representation is very useful for analyzing

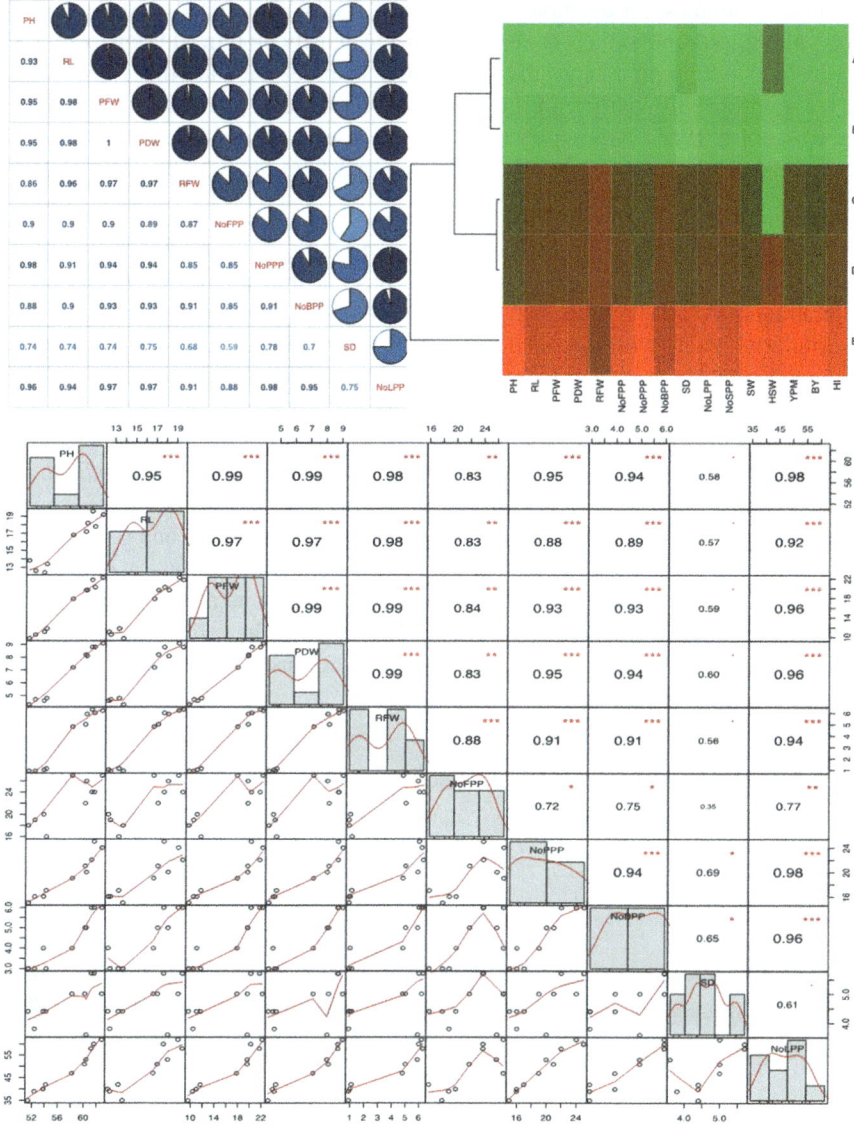

Fig. 5.2 Correlogram (top left), heatmap (top right), and scatterplot (bottom) of an agricultural data set based on chickpea (*Cicer arietinum*) growth in different soil moisture. *PH* plant height, *RL* root length, *PFW* plant fresh weight, *PDW* plant dry weight, *RFW* root fresh weight, *NoFPP* number of flowers per plant, *NoPPP* number of pods per plant, *NoBPP* number of branches per plant, *SD* stem diameter, *NoLPP* number of leaves per plant. For treatments A, B, C, D, and E, see Methodology. *Packages*: corrplot (Wei and Simko 2017), PerformanceAnalytics (Peterson and Carl 2018); *Functions*: coorplot, heatmap, chart.Correlation

variation in samples, for example, the degree of dispersion (spread) and skewness. Violin plot is a combination between box plot and density plot. In addition to all box plot features, violin plot shows the distribution shape of the data, that is, showing the kernel probability density of the data at different values, similar to histograms (see Rosenblatt 1956). Violin plot is specifically useful to see if the data is having multimodal (more than one peak) distribution or where the data have points that are more frequent.

Example The box plot (Fig. 5.3, left) shows variation in plant height among the treatments. In a general, we can observe that chickpea individuals that were seeded in treatments A and B were larger, with slightly higher values for treatment A. Similarly, plants with intermediate growth for treatments C and D and plants with low height for treatment E are indicated. The box plot with "notch" (Fig. 5.3, middle) is similar to box plot (Fig. 5.3, left), but it shows the confidence interval. Usually if we have overlap of notches between two groups, it means that there is no difference between them. Thus, comparing the treatments in Fig. 5.3 with notch, no differences between A and B or between C and D are observed. However, it may be assumed that there is a difference between A–B and C–D, A–B and E, and C–D and E. Furthermore, the violin plot (Fig. 5.3, right) shows the distribution of the data based on the kernel probability density. Analyzing the treatment B, we can see that there is no multimodal distribution of data, where most of the values are distributed around the median. On the other hand, treatment C, for instance, presents two peaks of distribution, showing a bimodal distribution.

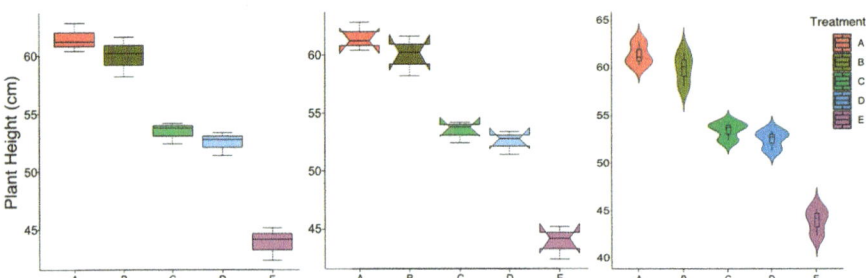

Fig. 5.3 Box plot (left), box plot with notch (middle), and violin plot (right) of an agricultural data set based on chickpea (*Cicer arietinum*) growth in different soil moisture. *PH* plant height, *RL* root length, *PFW* plant fresh weight, *PDW* plant dry weight, *RFW* root fresh weight, *NoFPP* number of flowers per plant, *NoPPP* number of pods per plant, *NoBPP* number of branches per plant, *SD* stem diameter, *NoLPP* number of leaves per plant. For treatments A, B, C, D, and E, see Methodology. *Package*: ggplot2 (Wickham 2016); *Function*: ggplot

5.6   Chord Diagram and Bipartite Networks

Overview Chord diagram is a graphical method to visualize the inter relationships between groups in a matrix, and it is very useful to see and compare connections, similarities, and differences among groups. Lines or arcs link one group to the other, and the width of the arc is proportional to the "importance" of the flow (if weighted). In a similar way, Bipartite Networks show connections (links) among nodes from two distinct sets and can be binary (presence/absence) or quantitative (weighted). In addition, different metrics (e.g., nestedness, connectance) can be used to evaluate the network. For this, one can use the *network-level* function in package "bipartite."

Example The chord diagram (Fig. 5.4, left) shows that the majority of bands present a high number of bindings with genotypes, except B1 and B2, which have links

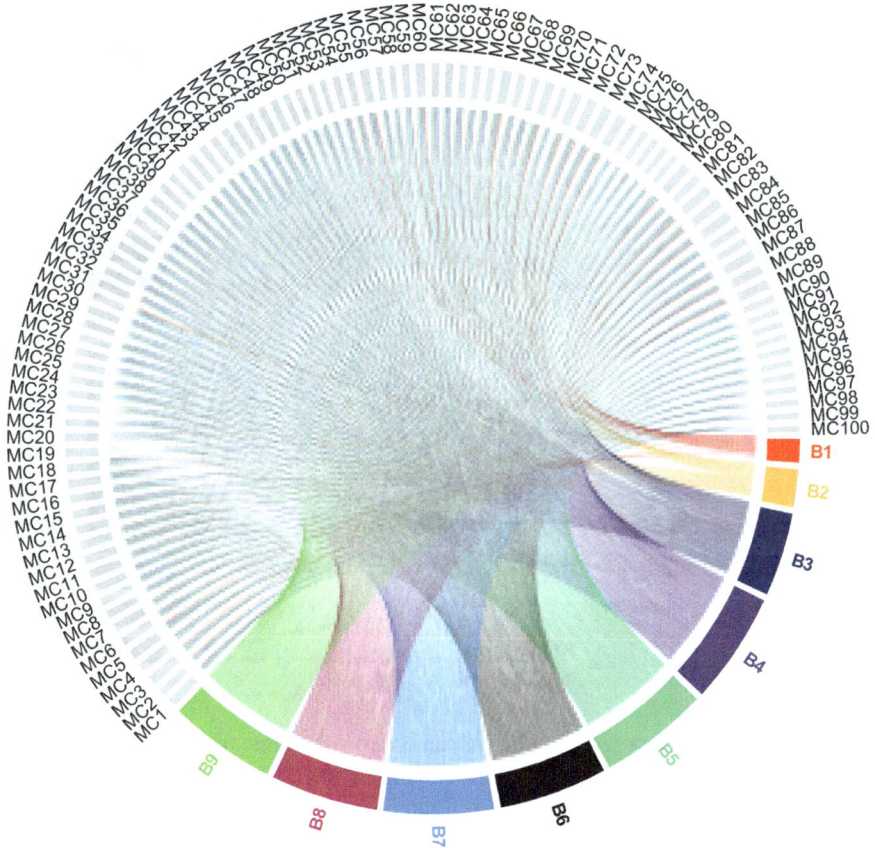

Fig. 5.4 Chord diagram showing the relationship between bands (B) and genotypes (MC)

Fig. 5.5 Bipartite network showing the relationship between bands and genotypes in *plant species*. *B* bands (orange), *MC* genotype (blue). Observe the decreasing order of bands in relation to the number of connections with the genotypes. *Package*: circlize (Gu et al. 2014), bipartite (Dormann et al. 2008); *Functions*: chord diagram, plotweb

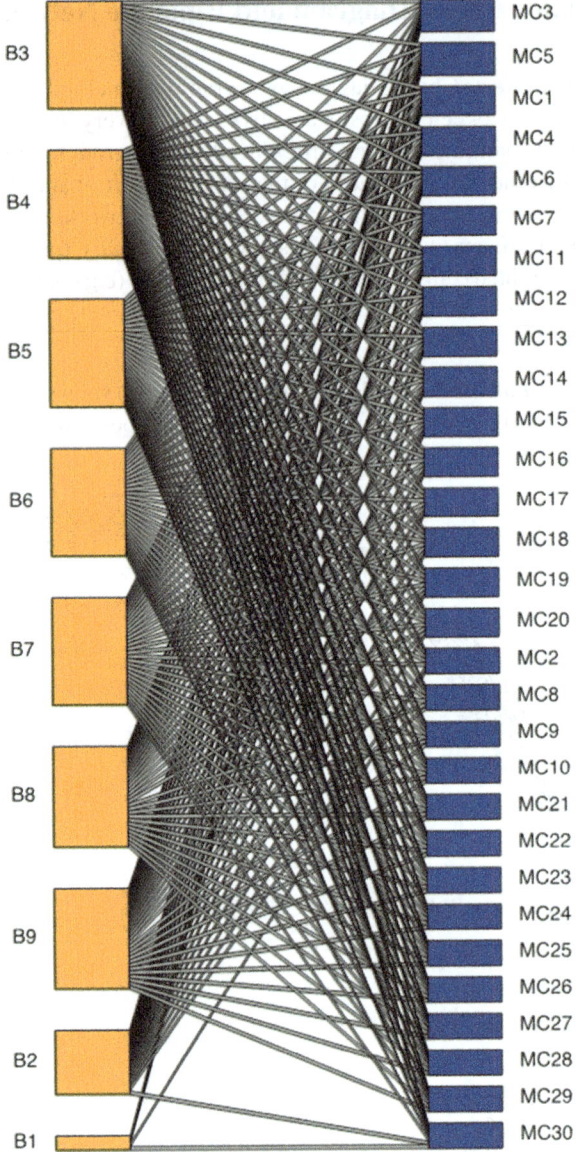

with only a few genotypes. Similarly, we can observe the same pattern of interactions in the bipartite network (Fig. 5.5). However, the network is more informative, since provides details of the decreasing order of interactions, showing the bands that have the greatest and the least number of interactions.

5.7   Hierarchical Clustering

Overview Hierarchical clustering or hierarchical cluster analysis is an algorithm that clusters similar objects into specific groups. These clusters have a predetermined ordering from top to bottom from distance measures, such as Euclidean distance.

Example The cluster dendrogram, based on Euclidean distance (Fig. 5.6), shows the grouping of treatments according to the characteristics of each group. In our case, three groups, A–B, C–D, and E are observed, which are in agreement with the graphics made in Sect. 5.5. Thus it may be suggested that there are differences between these three groups based on the Euclidean distance of the characteristics analyzed in the treatments.

5.8   Final Remarks

Many research studies in agricultural systems have been developed so far. However, the studies did not present adequate statistical approaches or graphical methods that explain and better represent the results obtained. This directly influences the visualization and dissemination of the works, and limits the scope and visibility of

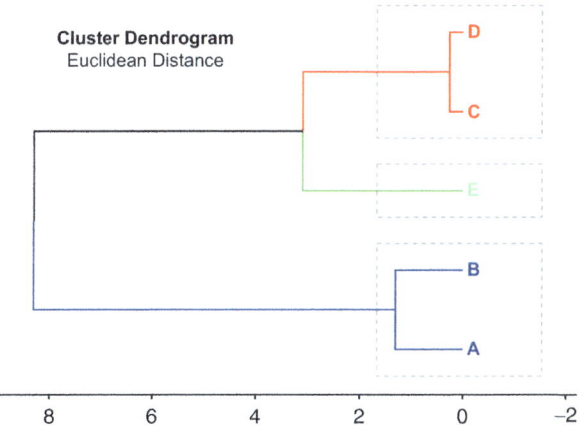

Fig. 5.6 Cluster dendrogram showing the grouping of treatments according to the characteristics of each group. Groups of different colors differ by the Euclidean distance applied in the analyzed characteristics of each treatment (see Sect. 5.5). Package: cluster (Maechler et al. 2018), factoextra (Kassambara and Mundt 2017); Functions: dist, hclust, fviz_dend

valuable contributions in magazines of high impact and potential. Therefore, the application of advanced statistical approaches is necessary and suggested so that researchers can better evaluate and present the results of different studies related to the agricultural system.

References

Dormann CF, Gruber B, Fründ J (2008) Introducing the bipartite package: analysing ecological networks. R News 8:8–11. https://doi.org/10.1159/000265935

Dunn CP, Stearns F (1987) Relationship of vegetation layers to soils in southeastern Wisconsin forested wetlands. The American Midland Naturalist 118:366–74.

Gu Z, Gu L, Eils R et al (2014) *circlize* implements and enhances circular visualization in R. Bioinformatics 30:2811–2812. https://doi.org/10.1093/bioinformatics/btu393

Kassambara A, Mundt F (2017) factoextra: extract and visualize the results of multivariate data analyses. R Packag. version 1.0.5.999

Maechler M, Rousseeuw P, Struyf A, et al (2018) Cluster: cluster analysis basics and extensions. R Packag. version 2.0.7-1

Peterson BG, Carl P (2018) PerformanceAnalytics: econometric tools for performance and risk analysis. https://cran.r-project.org/package=PerformanceAnalytics. Accessed 18 Oct 2018

Rosenblatt M (1956) Remarks on some nonparametric estimates of a density function. Ann Math Stat 27:832. 27:832. https://doi.org/10.1214/aos/1176348654

Tang Y, Horikoshi M, Li W (2016) Ggfortify: unified interface to visualize statistical result of popular R packages. R J 82:478–489

Wei T, Simko V (2017) R package "corrplot": visualization of a correlation matrix (Version 0.84)

Wickham H (2016) ggplot2: elegant graphics for data analysis. Springer-Verlag, New York

Chapter 6
Data Measurement, Data Redundancy, and Their Biological Relevance

Mohd Sayeed Akhtar, Ibrahim A. Alaraidh, and Mallappa Kumara Swamy

Contents

6.1 Introduction.. 103
6.2 Database Annotation Quality... 104
6.3 Database Redundancy.. 104
6.4 Genomes: Diversity, Size, and Structure... 105
6.5 Gene Content in Genomes... 105
6.6 Protein and Proteomics... 106
6.7 Protein Length, Distribution, and Function... 107
6.8 Conclusions and Future Prospects... 107
References... 107

6.1 Introduction

The fundamental feature of DNA and protein molecules is that they can be organized in the form of digital symbols and digitally stored as a data (De Silva and Ganegoda 2016). Nucleotides (adenine, guanine, thymine, and cytosine) and amino acids (tyrosine, glycine, histidine, lysine, etc.) are distinct, although chemically modified sometimes (Akhtar et al. 2017). Scientists are continuously stepping up their efforts to understand the genetic or biological processes that are connected clinically with the initiation and progression of various diseases. There is a flood of genomic and protein sequence data, which provides a clue on various biological processes, protein interactions, and disease paths. Thus, exploiting these sequence

M. S. Akhtar (✉)
Department of Botany, Gandhi Faiz-e-Aam College, Shahjahanpur, Uttar Pradesh, India

I. A. Alaraidh
Science College, Botany and Microbiology Department, King Saud University,
Riyadh, Saudi Arabia

M. K. Swamy
Department of Biotechnology, East West First Grade College of Science,
Bengaluru, Karnataka, India

© Springer Nature Switzerland AG 2019
K. R. Hakeem et al. (eds.), *Essentials of Bioinformatics, Volume III*,
https://doi.org/10.1007/978-3-030-19318-8_6

data can benefit from gaining new understanding. However, for clinical applications, the possible complications of digital data analysis have to be overcome. Biologists encounter some of the practical challenges, including data handling, data computing, and integrating different data resources, in the present genomic age. Therefore, experimentally determined sequences in principle have complete certainty. As a result, there is no limit of uncertainty associated with the efficiency of measurement. If we have enough economic resources, the nucleotide sequence in genomic DNA and the associated amino acid sequences can be revealed completely. However, in genomic projects that carry out large-scale sequencing, the purpose, relevance, ethics, and economics decide the quality of data. In this chapter, the fundamental aspects of database annotation quality, database redundancy, genomics diversity, and gene contents are discussed. The information will enable biologists and clinical experts to make a significant interpretation of a wide range of heterogeneous data for pharmaceutical applications.

6.2   Database Annotation Quality

The sequencing is the process of determining the accurate order of nucleotide bases such as A, G, C, and T in the strand of the DNA molecule. It is used to determine the sequence of individual genes, full chromosomes, or entire genomes of an organism. Although a lot of sequence data are determined for a range of different individuals accurately, all are not accessible to the researcher because of incorrect interpretation of experiments or incorrect handling and storage of public databases. The various reasons are as follows: (i) public databases are curetted by highly diverse people; (ii) they are annotated by highly diverse people; (iii) error rate of subsequent handling is more; (iv) there is some amount of experimental error; and (v) the way sequence or any biological data is stored in various biological databases.

A nucleus of plant cells and human brain both are supposed to handle a large amount of data accumulated throughout the life cycle of individual organisms, and the information is recalled by the content-addressable system, but the computers do not work in this fashion. A computer searches an address passport by name, profession, or hair color and needs perfect spelling. In contrast, a biological sequence search algorithm often has to use a fuzzy representation of its content. This is represented by numbers that indicate positions. A numeric representation cannot be reviewed by visual inspection. In sequence databases, the result is junk and random noise. It would mess up coding and noncoding regions, and the bioinformaticians should only take into account these potential sources of errors.

6.3   Database Redundancy

The most important problem in the biological data application is database redundancy. Many entries in DNA and protein databases represent members of the same family or versions of homologous genes found in different organisms.

Several groups might have submitted the same sequence. Annotation of the similar sequences would show close to identical results, but there may be significant differences between species and tissues. Redundancy may also result from different experimental approaches. In the same gene, this may produce variation. In a large part of c-DNA splicing, the spliced form of pre-mRNA means that genes are undergoing alternate splicing. A given piece of DNA may be associated with several c-DNA, which would lead to sequence discontinuous with genomic DNA. As a result, there would be different ways of joining the coding and noncoding regions. Data redundancy plays a nontrivial role in massive parallel gene expression. The sequence of genes being spotted on glass plates or synthesized on DNA chip is typically based on the sequences or clusters deposited in databases. Microarray would end up in sequences more than the number of genes.

Biological data may also represent protein sequences. It is possible that protein sequence may not directly correlate with genomic wild-type sequence due to modifications or requirement for crystallization. A redundant data set may result in (i) biased statistical analysis and (ii) correlation between different positions, which may be an artifact of biased sampling. Predictions or calibrations of too closely related data sets may be wrong. However, one should be very careful in attempting to discard valuable information of too closely related sequences. One way to avoid redundancy is to give weightage based on novelty. Another way is to develop a sequence profile, i.e., position-by-position amino acid variation.

6.4 Genomes: Diversity, Size, and Structure

Genome differs in size and storage principle (DNA or RNA) and may be single- or double- stranded. Cellular genomes are made up of DNA, while phage and viruses may be single- or double-stranded DNA or RNA. Bacteriophage $\Phi \times 174$ was the first genome to be sequenced, having 5386 base pairs (Table 6.1). Small genomes usually come in one piece. However, archaean *Methanocaldococcus jannaschii* completed in 1996 has several chromosomes. In chimpanzees, there are 23 chromosomes in addition to the 2 sex chromosomes. Similarly, cauliflower mosaic virus has 7 genes and the yeast has 5800 genes.

6.5 Gene Content in Genomes

Genes are one or several segments that constitute expressed units. It is revealed from the human genome project that the number of genes is quite small (30,000). The fruit fly has 14,000 genes. The mass of nuclear DNA in the haploid genome is called a C-value. The number of bases in gene bank for a most sequenced organism is tabulated in table form (Table 6.2). The chromosomal sizes do not appear as the original order of chromosomes. The total reference human genome sequence seems to contain 11,953,879,540 base pairs, while the numbers of sequence bases are

Table 6.1 Estimated number of genes and genome sizes in different organisms

Organisms	Genes	Genome size (in Mbp)
Bacillus subtilis	3700	4.200
Bacteriophage	270	0.166
Cauliflower mosaic virus	7	0.008
Escherichia coli	4243	4.600
HIV type 2	9	0.009
Homo sapiens	30,000–40,000	3100
Methanocaldococcus jannaschii	1729	1.66
Mycoplasma genitalium	477	0.580
Mycoplasma pneumoniae	716	0.820
Mus musculus	30,000	3300
Myxococcus xanthus	7311	9.100
Saccharomyces cerevisiae	5800	12.10

Table 6.2 Number of bases in gene bank for the most sequenced organisms

Species	Number of bases	Number of entries
Arabidopsis thaliana	648,235,477	9,81,930
Brassica oleracea	404,132,383	5,96,255
Gallus gallus	684,548,519	7,87,971
Homo sapiens	11,953,879,540	98,99,176
Macaca mulatta	434,454,818	68,350
Medicago truncatula	420,276,295	3,87,953
Mus musculus	7,917,536,708	66,25,240
Oryza sativa	1,171,720,224	3,57,923
Sorghum bicolor	463,997,658	7,84,734
Triticum aestivum	349,010,415	6,12,712
Xenopus tropicalis	954,916,818	11,79,107
Zea mays	1,780,527,217	26,78,230

1,780,527,217 and 1,171,720,224 for maize and rice plants, respectively. However, it is estimated as 648,235,477 bases for *Arabidopsis thaliana*.

6.6   Protein and Proteomics

At the level of protein, large-scale analysis of complete genome is known as proteome analysis. It is the total protein expression as a number of chromosomes in multicellular organisms. It may change with the cell type and time. The protein undergoes glycosylation and phosphorylation. There are many other types of posttranslational modifications such as addition of fatty acids and cleavage of signal peptides in the N-terminal of secretory proteins. These are experimentally determined examples which are deposited in public databases. The prediction of posttranslational modifications in proteins is an area of interest for bioinformaticians.

6.7 Protein Length, Distribution, and Function

Statistical analysis plays a vital role in the evolution of protein sequences, especially the local and nonrandom patterns. By using the data on soluble proteins, one can formulate 10,112 natural sequences with 100 amino acids. The functional aspect is determined mainly by local characteristics. It does not depend critically on the structure maintained in long-range order. Functionally identical proteins are evolutionary related or homologous and have similar folds. A typical example is serine proteases (Geer 1981). However, it is possible to conserve the function when the sequence is not evolutionary related, e.g., transaldolase fructose-1,6-bisphosphate aldolase, urease catalytic domain, and phosphotriesterase. Enterotoxin and chlorera toxins are closely homologous. The fold is similar to TSS toxins, which is remote homology. However, aminoacyl tRNA synthetase has the same fold, but 4.4% homology is not understood (Russell et al. 1997).

6.8 Conclusions and Future Prospects

The exploitation of genomic and proteomic data sequence benefited us in multiple ways. The clinical observations of these digital data analysis have solved the challenges, including data handling and computing and integration. The experimentally determined sequences have complete certainty and efficiency in data measurement. Therefore, the future researchers will be more focused on the large-scale sequencing, relevance, ethics, and quality of data.

Acknowledgments The authors (Mohd Sayeed Akhtar and Ibrahim A. Alaraidh) are highly grateful to the Department of Botany, Gandhi Faiz-e-Aam College, Shahjahanpur, U.P., India, and Botany and Microbiology Department, Science College, King Saud University, Riyadh, Kingdom of Saudi Arabia.

References

Akhtar MS, Swamy MK, Alaraidh IA, Panwar J (2017) Genomic data resources and data mining. In: Hakeem KR, Malik A, Sukan FV, Ozturk M (eds) Plant bioinformatics. Springer International Publishing, Cham, pp 267–278

De Silva PY, Ganegoda GU (2016) New trends of digital data storage in DNA. Biomed Res 2016:8072463. https://doi.org/10.1155/2016/8072463

Geer J (1981) Comparative model building of the mammalian serine proteases. J Mol Biol 153:1027–1042

Russell RB, Saqi MAS, Sayle RA, Bates PA, Sternberg MJE (1997) Recognition of analogous and homologous protein fold: analysis of sequence and structure conservation. J Mol Biol 269:423–439

Chapter 7
Metabolomic Approaches in Plant Research

Ayesha T. Tahir, Qaiser Fatmi, Asia Nosheen, Mahrukh Imtiaz, and Salma Khan

Contents

7.1   Introduction.. 110
7.2   Plant Metabolites and Their Types.. 111
    7.2.1   Primary Metabolites.. 111
    7.2.2   Secondary Metabolites... 112
7.3   Metabolomics... 113
    7.3.1   Metabolic Profiling (Nontargeted Metabolomics).............................. 113
    7.3.2   Metabolite Target Analysis (Targeted Metabolomics).......................... 114
    7.3.3   Metabolic Fingerprinting.. 115
7.4   Environmental Metabolomics.. 115
7.5   Plant Metabolome Data Processing.. 116
    7.5.1   Metabolomics Databases... 116
7.6   Data Handling.. 119
    7.6.1   Data Processing and Pre-treatment... 119
    7.6.2   Data Analysis... 122
7.7   Analytical Techniques in Metabolomics... 125
    7.7.1   Mass Spectrometry-Based Methods (MS).. 125
    7.7.2   Nuclear Magnetic Resonance (NMR) Spectroscopy............................. 129
    7.7.3   Vibrational Spectroscopy.. 130
7.8   Metabolomics: The Apogee in Realm of OMICS... 130
7.9   Challenges and Perspective in Agricultural Metabolomics.............................. 132
References... 134

A. T. Tahir (✉) · Q. Fatmi · A. Nosheen · M. Imtiaz · S. Khan
Department of Biosciences, COMSATS Institute of Information Technology,
Islamabad, Pakistan
e-mail: ayesha.tahir@comsats.edu.pk

© Springer Nature Switzerland AG 2019
K. R. Hakeem et al. (eds.), *Essentials of Bioinformatics, Volume III*,
https://doi.org/10.1007/978-3-030-19318-8_7

7.1 Introduction

Plant metabolomics is a relatively new and emerging field in parallel with advances in genomics, transcriptomics, epigenomics, proteomics, and phenomics. In an era of these rapidly evolving "omics" fields, crop metabolomics was among the list of neglected domains particularly for agronomic trait mapping and selection of plants (Parry and Hawksford 2012; Kumar et al. 2017).

Plant metabolites are complex in nature and have huge diversity, varying in number from 200,000 to 1,000,000 (Fiehn 2002; Saito and Matsuda 2010; Afendi et al. 2012). However, only a small percentage of diversity is known; among 350,000 plant species, only 15% have been characterized for their chemical/biochemical signatures (Cragg and Newman 2013). Yet, this metabolomics potential could be beneficial to decipher complex and superior traits for plant breeding (Hegeman et al. 2010; Zivy et al. 2015).

Collaborative efforts among the international plant metabolomics scientific community were boosted after an "International Plant Metabolomics Congress" held in 2002, at Wageningen, the Netherlands. A website (www.metabolomics.nl) was developed for access of abstracts and papers and is still functional for future collaborations and updates in plant metabolomics. Since then, many databases, tools, and web servers are available with the help of in silico efforts, for metabolomics analysis (Fukushima and Kusano 2013). Bioinformatics is playing a central role in system biology approaches, since computer-aided data systems were already integrated with mass spectrometers in the 1970s (Meier et al. 2017). Thereafter, rapid advances led to coupling of digital mass spectra with MS instruments (such as GC or LC) for fine separation of small molecules. The most significant task for any metabolomics experiment is interpretation of biological data to gain meaningful information about metabolites (Johnson and Lange 2015). Computational metabolomics is becoming popular with involvement of bioinformatics tools for carrying out statistical analyses and designing tools for aid in data processing to speed up workflow, developing metabolomics platforms and databases.

Computer-aided identification, screening and minimizing feature redundancy, workflow automation, meta-analysis for deconvulation of feature list, and mapping of identified metabolites on their pathways are the major tasks of computational metabolomics at present. They are extending and becoming more complex but more comprehensive with generation of huge amounts of data from various "omics" platforms (Wanichthanarak et al. 2015; Wen et al. 2018). Various analytical platforms are being used for analysis of metabolites, giving choices for metabolite target analysis to metabolome profiling, and can cover both known and unknown metabolites (Dettmer et al. 2007; Zhang et al. 2012; Jorge et al. 2016). However, single experiment cannot uncover the complete metabolic composition of a plant sample. Integrated strategies from sample preparation to its quantitation coupled with other "omics" are the key to unlock this complex world.

Recent advances in analytical and computational tools for metabolomics as well as computational tools developed for integration of metabolomic with other omics (particularly genomics) are discussed in the present chapter. In addition, we discuss their importance in model as well as cultivated crop species, and we also highlight the challenges and future prospectives of present-day agricultural metabolomics.

7.2   Plant Metabolites and Their Types

Plants have been used as sources of food and medicine since the beginning of civilization. These nutritional and medicinal properties are due to the presence of a wide variety of compounds which are known as metabolites (Gupta et al. 2017). Plants are sedentary organisms and are constantly interacting with a huge variety of potentially detrimental abiotic and biotic factors in their habitat. Hence, complex defense mechanisms are needed for their survival within the ecosystem. Among these mechanisms is their inherent immune system which is a type of chemical defense to cope with the hostile environmental conditions. This innate system of defense, which is of adaptive significance for the plants to survive in diverse ecological niches, creates a rich repertoire of complex compounds called metabolites (Ncube and Staden 2015).

Metabolites are the by-products of metabolism and occur naturally within the cells and are the products of enzyme-catalyzed reactions. For a compound to be classified as a metabolite, it must meet some specific criteria. These criteria include the following: (a) the compound must be present inside the cell, (b) it can be followed up by the enzymes to enter further reactions, and (c) it must assist some beneficial biological reactions taking place inside cells (Herwig and Ludwig-Muller 2014).

Enzyme-catalyzed reactions known as metabolic pathways result in these small, organic compounds, naturally found in plants (Edward 2014). Metabolites can be primary or secondary depending on their function inside the plant. Primary metabolites are crucial for essential plant functions such as growth and development, while secondary metabolites have specific functions (Paupière et al. 2014; Kumar et al. 2015). These are the compounds that help in the interaction of plants with other organisms. These include plant-pollinator, plant-pathogen, plant-herbivore interactions, etc. One of the best properties of metabolites is that they have a limited survival in the cell and they do not accumulate in the cell.

7.2.1   Primary Metabolites

These are the heavy-molecular-weight compounds which include metabolites, such as lipids, carbohydrates, proteins, amino acids and nucleic acids, which are synthesized by the cell via central metabolisms such as glycolysis, Kreb's cycle, and Kelvin cycle. In plants, these are results of essential metabolic pathways and so are termed as essential metabolites (Caretto et al. 2015).

Primary metabolites are important for plant survival as they are directly involved in normal growth, development, and reproduction. They also play an important role in normal physiological processes, synthesis assimilation, and degradation of organic compounds (Aharoni and Galili 2011; Wurtzel and Kutchan 2016).

Although primary metabolites are generally extracted using methanol-water-based solutions, various precautions are necessary to avoid possible artifacts. Contamination with natural metabolites of biological systems and storage temperature of extracted metabolites before their measurement are the two critical factors to consider for accurate measurements (Sauerschnig et al. 2018).

7.2.2   Secondary Metabolites

Secondary metabolites are biosynthetically derived from primary metabolites (Fig. 7.1) and are beneficial for plants in various aspects such as defense chemicals and detoxifying agents (Ncube and Staden 2015; Shitan 2016). Secondary metabolites are gaining attention as potential sources of beneficial contributors for cure of various diseases in humans as well (Chandra et al. 2017). Examples include alkaloids found in the plants of family Amaryllidaceae, e.g., galanthamine, which not only exhibit antimalarial, antitumor, antiviral, and immune-stimulatory activities but also could help in the cure of Alzheimer's disease (Conforti et al. 2010; Castillo-Ordóñez et al. 2017).

In spite of their importance in plants for several benefits, their participation in normal plant growth/development is limited to specific taxonomic groups (Noel et al. 2005; Ncube et al. 2012). More importantly they not only regulate communication between plants and their biotic or abiotic environment but also arbitrate many physiological aspects of growth and development, reproduction, and symbiosis and are also important constituents of secondary cell walls in the form of lignin (Brown et al. 2001). They are produced by pathways that are not common (unlike primary

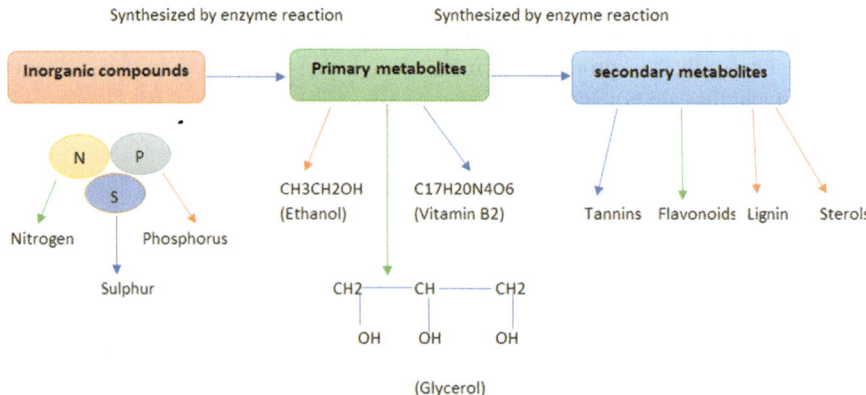

Fig. 7.1 Metabolite synthesis: primary to secondary metabolite production

metabolites) to all plants and have a limited survival in the cell, being not only specific to plant species but also specific to a plant developmental stage.

Interestingly, secondary metabolites are diverse in their structure, but their synthesis is carried out by the primary metabolism products, which are conserved in plant species. Immense research in biochemistry, molecular biology, and genomics elucidates that the diversification in the few central intermediates is the cause of secondary metabolites specificity at various levels. It involves the diversification of enzymatic pathways and enzyme families from substrate to products (Hartmann et al. 2005; Ncube et al. 2012). Synthesis of secondary metabolites also needs many transcriptional factors which are involved in the activation and concurrent expression of primary metabolism pathway genes. For example, in *Arabidopsis thaliana*, the regulation of glucosinolate biosynthesis needs not only the metabolic space but also the metabolic network from the primary metabolism. By the overexpression of two clades of genes such as ATR1-like and MYB28-like genes, the regulation of aliphatic and indole glucosinolate biosynthesis pathways is carried out, and as a result, the induced genes carried out assimilation of sulfur but also induce the formation of both glucosinolate molecules (Malitsky et al. 2008). These metabolites are grouped into three classes based on their structures and biosynthetic origins; these include terpenoids, phenolics, and alkaloids.

7.3 Metabolomics

Metabolomics is a term that is used for the quantification and identification of small molecules known as metabolites of the biological organization such as cells, tissues, organs, organic fluids, and organisms at a certain period of time (Daviss and Bennett 2005). Metabolites are the end products of metabolism; more precisely any molecule whose size is less than 1 kDa comes under this category (Samuelsson and Larsson 2008). The interaction of these metabolites with a biological system is known as metabolome (Jordan et al. 2009). To measure the whole metabolome with a single extraction and analytical tool is difficult because of the diverse nature of these molecules. The extraction method and analytical tools used in metabolomics, depends mainly on research objectives (Last et al. 2007). General experimental procedure used in plant metabolomics is illustrated in Fig. 7.2. Consequently analytical strategies could be (1) metabolic profiling (nontargeted or also called metabolomics), (2) metabolite target analysis (targeted), and (3) metabolite fingerprinting (Fiehn 2001; Nielsen and Oliver 2005; Patti et al. 2012).

7.3.1 Metabolic Profiling (Nontargeted Metabolomics)

This approach aims to identify as many metabolites as possible to conduct comparative analyses among related predefined groups such as amino acids, carbohydrates, organic acids, etc. The nontargeted metabolomics provides a complete examination

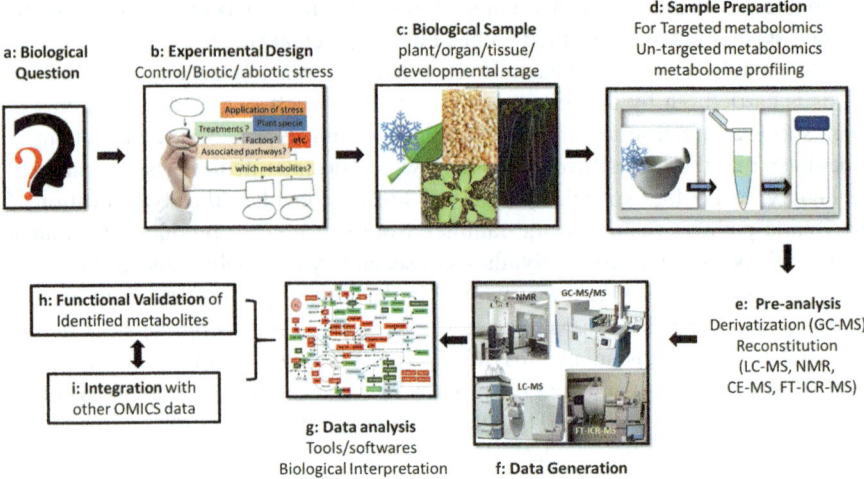

Fig. 7.2 Metabolomics workflow for high-throughput data analysis

of all the quantifiable analytes in a sample, including those that are novel (Daskalchuk et al. 2006). For unknown/novel metabolites, different experiments can be carried out with functional assays to check the novel metabolite for its biological activity and role, and finally addition to a database (Vinayavekhin and Saghatelian 2010).

Due to the inclusive nature of nontargeted metabolomics, this approach must be combined with some advanced techniques such as the examination of multivariate in order to divide the wide sets of data produced into smaller sets of convenient signals. Annotations such as in silico libraries or experimental studies are required with these manageable signals for the identification of metabolites. In addition this system could also help in fast processing of data, which indeed is a big challenge for this approach (Roberts et al. 2012).

7.3.2 Metabolite Target Analysis (Targeted Metabolomics)

Targeted metabolomics is a method in which a class of known metabolites is quantified. These methods need precision, rapid throughput, and consistency. Through the procedure of internal standards, examination can be assumed in a quantifiable or semi-quantifiable way. This method takes advantage of the complete understanding of a massive collection of metabolic enzymes, the kinetics of these metabolic enzymes, the end products of these enzymes, as well as the biological pathways that are known for these enzymes to which they contribute. When using targeted metabolomics, the preparation of sample needs to be optimized, in order to reduce the supremacy of high-abundance metabolites, as all the examined species will evidently be demarcated and rational archaeology will not be passed through to downstream examination (Roberts et al. 2012).

The key point which can lead to experimental accomplishment in targeted metabolomics is analytical reproducibility. If analytical reproducibility is high, results can be considered as due to certain biological alterations. It also reduces the number of experimental replicates necessary (Zhou et al. 2016).

The two most important advantages of targeted metabolomics are:

- An approach that can be used to examine very large sets of samples, reducing the time necessary for the study.
- The exact identity of each molecule is determined with the aid of multiple reaction monitoring (MRM) examination. These kinds of MS examinations are made with the assistance of a tool called as triple quadrupole (George 2005).

7.3.3   Metabolic Fingerprinting

This approach does not intend to quantify individual metabolites but rather to provide a broadscale comparison of results. This conceptual procedure requires a fingerprint of all measureable chemicals which are necessary for comparison of samples and for differentiation analysis. For optimal data acquisition, many instruments (see Sect. 7.7) are in use depending on experimental strategy (Krishnan et al. 2005; Scott et al. 2010).

At present, with advancement of technology, a huge amount of data is available in data banks. Previously, in the 1990s, Henry Nix's famous statement depicted the situation as below (Nix 1990):

> Data does not equal information; information does not equal knowledge; and, most importantly of all, knowledge does not equal wisdom. We have oceans of data, rivers of information, small puddles of knowledge, and the odd drop of wisdom.

This challenge is becoming more and more serious day by day. Metabolomics is also facing the similar challenge of making sense of data produced from metabolomics centers. If we compare the above three strategies, targeted metabolomics is less challenging since metabolites are preselected, have high detection rate (as they are known), and are easy to quantify (Sawada and Hirai 2013; Mahdavi et al. 2015). For quantification, we can use either standard solutions for external calibration or internal standards for internal calibration. On the contrary, as untargeted metabolomics deals with both known and unknown metabolites, data operation is more complex and requires statistical and bioinformatics tools (see Sects. 7.5 and 7.6) to make the process less intensive, less time-consuming, and more efficient compared to manual data handling.

7.4   Environmental Metabolomics

Environmental metabolomics is also one of the emerging applications of metabolomics to study the interactions of organisms with their environment. This approach also helps in assessing the function and health of an organism at the molecular level

(Bundy et al. 2008). Metabolomics also have the wide range of applications in the environmental sciences, ranging from considerate response of organisms to abiotic stresses, in order to examine the responses of organisms to other biota (Bundy et al. 2008; Jorge et al. 2016). This approach is attained by determining the numerous small molecules within an organism, cell, tissue, or fluid, which can give the researchers a complete picture of their efficient metabolic phenotype (Viant 2008).

Environmental science is considered as an "easygoing" science as these studies are being accompanied with the most innovative instruments that are available nowadays. During the past few decades, research on environmental metabolomics has been carried out on a number of species, including freshwater and sea fishes and various species of aquatic and terrestrial invertebrates, as well as on plants and microorganisms. This approach has covered ideal or non-ideal organisms as well as the research laboratory and the field-based examinations.

7.5 Plant Metabolome Data Processing

7.5.1 Metabolomics Databases

A database is the collection of interrelated information or data on a specified subject that are stored in an organized way. Generally, a database is accessed and analyzed by the end users through a computer software application termed as database management system (DBMS). DBMS offers various features and functionalities that help in the administration and control of database such as creation, modification, and deletion of information, leading to an easy access and efficient data retrieval scheme for further analysis by a program. Some of the computer languages for DBMS include Oracle, MySQL, dBase, FoxPro, MS Access, etc.

A good database plays a key role in the improvement of all aspects of modern life. A database for science and education has a great impact on business as well. Understanding and extraction of useful information from the enormous amount of data produced by advanced technologies in the field of metabolomics necessitates the development of metabolomics databases. Below are few available metabolomics databases that help in metabolomics research:

- *Plantmetabolomics.org*
  Data related to plant metabolomics can be easily searched, visualized, and downloaded via the Plantmetabolomics.org web portal and database. This database assimilates data collected from multiple platforms and laboratories along with their comparative analysis using visualization tools. Each metabolite in Plantmetabolomics.org is associated with relevant experimental data from multiple annotation databases. Furthermore, this web portal also provides detailed tutorials on plant metabolomics experiments (Bais et al. 2010).

- *Plant Metabolome Database (PMDB)*
  Secondary metabolites of plants in 3D structures, available on other structural databases, are integrated in Plant Metabolome Database. It has a user-friendly interface along with integration of JME editor for visualization of metabolites in graphical as well as documented formats. PMDB also contains internal and external links to various relevant databases, i.e., KEGG, PubChem, and CAS Number (Udayakumar et al. 2012).
- *KEGG Pathways*
  One of the most popular and widely used databases, containing metabolic pathways from an extensive range of organisms, is known as KEGG (Kyoto Encyclopedia of Genes and Genomes). These pathways are hyperlinked to metabolite and protein/enzyme information (Kanehisa and Goto 2000).
- *KNApSAcK*
  KNApSAcK is a compound database that represents the correlation of metabolites with taxonomic information of different species. This database integrates information of reported metabolites from various organisms, particularly focusing on plants (Afendi et al. 2012).
- *Plant Metabolic Network (PMN)*
  Plant Metabolic Network (PMN) comprehensively integrates individual species-/taxon-specific databases which comprise of enzyme, pathways, and other information. PlantCyc database is one of the PMN's databases that indexes computationally prophesied and experimentally supported metabolic pathways and enzymes in over 350 plant species. The website can be accessed at https://www.plantcyc.org.
- *METLIN*
  METLIN is a comprehensive high-resolution tandem mass spectrometry (MS/MS) database which integrates over a million compounds including plant and bacteria metabolites as well as steroids, lipids, carbohydrates, central carbon metabolites, small peptides, and toxicants. Empirical and in silico MS/MS data has been calculated by the individual analysis of each metabolite and small molecules. This database also provides the facility of multiple searching options like batch, precursor ion, mass, or fragment searches (Smith et al. 2005).
- *NMR Metabolomics Database of Linkoping (NMR-MDL)*
  NMR-MDL (Metabolomics Database of Linkoping) is a freely accessible data repository that is devoted to the omics of small biomolecules. Metabolites that portray the primary cellular metabolism in various animals, plants, fungi, etc. comprise of the major part of this database. However, some of the metabolites that depict secondary cellular metabolism are also included in minority. Moreover, NMR-MDL also facilitates its users to access NMR parameters of metabolites particularly in aqueous phase (Lundberg et al. 2005).
- *Madison-Qingdao Metabolomics Consortium Database (MMCD)*
  Madison-Qingdao Metabolomics Consortium Database (MMCD) integrates metabolites from various biological samples that are identified and quantified using NMR and MS approaches. It also gathers information concerning small molecules of interest from scientific literature and other databases. The information

for each metabolite in MMCD is described in 50 separate data fields, i.e., chemical formula, synonyms, structure, NMR and MS data, physical and chemical properties, etc. Links to other databases like KEGG and PubChem are also present on this platform (Cui et al. 2008).

- *Bio-MassBank and MassBank*
  This depository encompasses a mass spectrum of identified as well as unidentified metabolites from plant and microbial samples through LC-MS, GC-MS, and CE-MS analyses. The metabolites identified could help to discover the existence of unidentified metabolites. Currently Bio-MassBank indexes approximately 664 entries from *Arabidopsis* leaf and 636 entries of *Lotus japonicus* flower, analyzed by LC-MS and MS/MS, respectively (Horai et al. 2010).

- *KomicMarket (Kazusa Omics Data Market)*
  Annotations of metabolite peaks detected by mass spectrometers are cited in the KomicMarket database. These peaks were obtained from high-resolution liquid chromatography MS. Until 2013, 75 samples from 10 plant species and 215 other standard chemicals were reported to be stated in this database (Iijima et al. 2008).

- *MassBase*
  MassBase is another database where experimentally known or predicted peaks found in biological samples are detailed. The database is also linked to other databases MassBank and KNApSAcK. The website of database is accessible at http://webs2.kazusa.or.jp/massbase.

- *Metabolome Activity DB*
  This database includes data about the activity of metabolites, i.e., how they may affect other organisms. The data describes affected organisms and the mode of action of metabolites in these organisms. The website of database is accessible at http://kanaya.naist.jp/MetaboliteActivity/top.jsp.

- *Metabolomics.jp*
  Metabolomics data is maintained and curated in the Metabolomics.jp database by Arita Lab at the University of Tokyo. This web portal contains the databases comprising data related to drugs, basic metabolites, flavonoids, and plant taxa. The wiki is accessible at http://metabolomics.jp/wiki/Main_Page.

- *ReSpect*
  ReSpect is a plant metabolite-specific MS2 database. All the spectral records in this database are annotated with taxonomic information, i.e., from where that metabolite was extracted and to which structural class it belongs (Sawada et al. 2012).

- *ChemDP*
  ChemDP is a natural product-based chemical database of Pakistan that contains high-quality 3D structure of plant metabolites isolated and elucidated from X-ray, NMR, MS, and other techniques. The database offers ready-to-use structures in various file formats for computational docking and simulation studies. The database currently contains over 1000 compounds along with their reported bioassay data (Mirza et al. 2015).

7.6   Data Handling

Metabolomics is an extensive study of metabolite profiling, with the following work-flow: sample preparation, data processing and pre-treatment, data analysis, metabolite identification, and interpretation of the data generated (Defernez and Le Gall 2013). A colossal range of data is produced by metabolomic experiments; therefore, handling and analyzing the data have a huge impact on the identification and quantification of particular metabolites (Boccard et al. 2010). In this topic, strategies for handling and analyzing metabolomic data are discussed as given in Fig. 7.3.

7.6.1   Data Processing and Pre-treatment

The phrase "garbage-in, garbage-out" is more relevant to the era of undeniably advanced digital and computer technology, where the fast and powerful computers can produce a large amount of erroneous data or information in a short period of

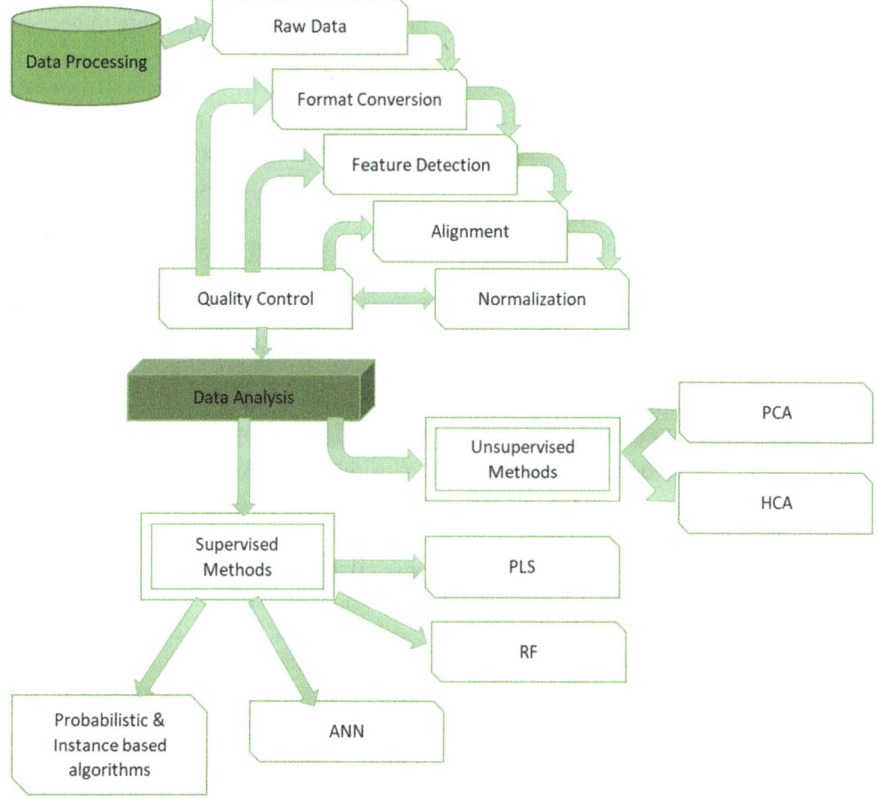

Fig. 7.3 Schematic representation of steps involved in metabolomics data processing and methods used in data analysis

time due to flawed input or inexperienced technologists. Therefore, the data pretreatment or preprocessing is fundamentally the most important step in data mining and data analysis. The following are some methods that are used in pre-treatment of raw data for metabolomics profiling:

- *Data Format Conversion*

  File format conversion of collected raw data is the initial step in data processing. Commonly, the mass spectrometry (MS) vendor-dependent binary format is converted into other common formats that enable independent software and operating systems to carry out processing. Hyphenated mass spectrometry-based data is commonly saved in NetCDF and mzXML file formats. These common formats allow mutual sharing of data among laboratories. Owing to advances in resolution and throughput of MS data, the resulting data production has increased exponentially, causing difficulty in handling such enormous datasets. The easiest way to reduce data size is to apply irreversible filtering methods that remove small intensity peaks and compressions, as applied in open-source software for mass spectrometry data processing like *mzMine* and *mzMine2*; however, it comes with a severe shortcoming of distorting data analysis. Some other MS data analysis software like *Mass++* facilitate direct import of binary files provided by MS vendors into the software; however, these are vulnerable to access binary data files from application programming interface (API). Moreover, most of the hyphenated MS binary data files, e.g., .wiff files or .D formats, contain mass spectra data, and if the points in the mass spectra are not unique on the chromatograph, then this kind of data structure takes much longer to yield a chromatograph or electropherogram. However, despite a trade-off between quick data access and accessibility of a standard file format, there is a need for development of standardized file formats without the limitation of quick access (Sugimoto et al. 2012).

- *Feature Detection*

  Feature or peak detection is one of the important steps in data processing. The main objective of peak detection is to diminish false positive detection and to provide precise quantitative information of ion concentrations. True signals can be distinguished from noise in three following ways. In the first technique, the independent peak detection is carried out in two directions, namely, mass-to-charge ratio (m/z) and retention time. This method sets a threshold level in both directions based on all intensity values along the vector, and data points above the threshold level are called peaks. The second approach involves the examination of single ion chromatograms, where each chromatogram covers a small range of m/z; thus the problem of searching peaks in m/z direction is reduced. A threshold level of these chromatograms can be calculated based on mean or median of the chromatogram or by applying a Gaussian filter. The third way for feature detection is model fitting of a three-dimensional standard isotope to the raw signal followed by removing noise from the real signals. Background reduction of false peaks is conducted by repeated iterations of this procedure. Detection of entire isotope model pattern instead of single peaks can reduce the detection of false positive peaks (Katajamaa and Orešič 2007; Boccard et al. 2010).

*   *Alignment*
    The key procedure of data processing is the alignment of multiple datasets in both m/z and retention time dimensions to combine outputs. The purpose of alignment is to diminish migration shifts among datasets. For this purpose, multiple alignment techniques have been established (Sugimoto et al. 2012). Migration shifts along m/z axis are relatively easily reduced using advanced chromatographic techniques; however, minimizing shifts in retention times are more challenging. The alignment of algorithmic techniques requires a reference chromatogram to reduce retention time shifts. Several problems need to be addressed while performing reducing shifts. For instance, the choice of a reference chromatogram is of utmost importance, as it has a great impact on the results. Moreover, chemical selectivity must be preserved among samples of different composition, and it is expected that peaks should be shifted by reduced distance among adjacent peaks. Additionally, alignment algorithms need to be fast enough to align large datasets rapidly, without affecting the resultant quality of output.

    Generally, three approaches are used to align multiple chromatograms. The simplest approach introduces the binning of false data along the chromatographic dimension. In this way, none of the information is compromised, and all the false peaks or errors are summed back toward the boundaries. The second technique is automatic and needs minimal manual work. In this method, alignment is carried out by stretching the retention time axis of all samples to a standard reference. This approach is one of the most investigated alignment approaches (Boccard et al. 2010). Additionally, various warping algorithms are used as alternatives of this method, but two algorithms were initially established, i.e., "Time Correlation Optimized Warping" and "Dynamic Time Warping" (Nielsen et al. 1998; Pravdova et al. 2002). The Time Correlation Optimized Warping (COW) works by dividing chromatograms into smaller segments and increasing the correlation coefficient between reference and sample chromatograph by shifting each segment. In contrast, Dynamic Time Warping (DTW) detects matching peaks among multiple sample chromatographs. Previously, this algorithm was used for homology searching of genes and genomes. Unfortunately, both of these algorithms have a limitation, as they are only applicable to smaller datasets because they consume a great deal of time in aligning larger datasets (Sugimoto et al. 2012). Furthermore, detection of signals using curve resolution technique is another solution for matching corresponding peaks among samples. Various algorithms like GENTLE (Shen et al. 2001), MEND (Andreev et al. 2003), and progressive clustering (De Souza et al. 2006) are available to detect and rearrange the corresponding peaks of different samples by the use of time and mass/charge tolerance. Components with high spectral similarity can be matched for the assimilation of similar peaks (Boccard et al. 2010).

- *Normalization/Scaling*

  Normalization or scaling is an indispensable step, as it removes the unwanted systemic bias while upholding biological variation among datasets. It is a relatively challenging task due to chemical diversity of complex data, leading to incongruities in the ionization productivities. Deviation in signal strengths due to measurement errors must also be eliminated. Two strategies are commonly applicable for normalization of metabolomics data. Statistical models symbolize simple methods to evaluate scaling factors among metabolomics samples, i.e., scaling by unit norm or median intensities. The second mode to perform normalization is the addition of a set of internal (added to sample before extraction) and external (added to sample after extraction) standards that are based on specific regions of retention time (Katajamaa and Orešič 2007; Boccard et al. 2010).

- *Quality Control*

  Various algorithms and techniques have been developed for data processing and alignment. However, the selection of best algorithm is a difficult task. Therefore, quality control performs comparison tests to check the output of each algorithm. In a study, the feature detection algorithm for high-resolution LC-MS datasets, *centWave*, was compared with filters employed in *XCMS* and *MZmine* software, and the results indicated that the peaks obtained from these software did not overlap; rather some peaks were only detected by one software. Similarly, the alignment of LC-MS data was conducted on four different software, namely, *XCMS* (Smith et al. 2006), *MZmine* (Katajamaa et al. 2006), *msInspect* (Bellew et al. 2006), and *OpenMS* (Bertsch et al. 2011), and it was observed that no single software produced perfect alignment of the data. Therefore, several means are required to improve output quality. One of the solutions to improve performance is to develop an iterative framework with machine learning technology that would allow improvement of parameters to generate high-throughput results (Sugimoto et al. 2012).

7.6.2   Data Analysis

When the data matrix along with peaks is generated from raw data, various statistical analysis procedures are conducted in order to identify metabolites and their biological significance. Analytical procedure comprises of two phases. In the first phase, the multivariate analysis is used to generate an overview of the datasets followed by grading of individual peaks by univariable analysis in the second phase (Sugimoto et al. 2012). These analytical methods are also divided into two categories, namely, unsupervised and supervised methods.

7.6.2.1   Analysis by Unsupervised Methods

Unsupervised learning method is a type of machine learning algorithm that attempts to analyze a dataset without measuring any related response. Unsupervised learning helps to understand the relationship among the dataset partitions and also focuses on the variables accountable for these interactions. The following are the two types of clustering techniques that fall in this category:

- *Principle Component Analysis (PCA)*
  Principal component analysis (PCA) is the most extensively used multivariate, unsupervised statistical analysis technique. It was developed prior to other analytical techniques like pattern recognition, clustering, etc. and was used as the first analytical technique. Specific variables called the "Principle components" are defined in this technique, to describe the maximum variance in the distribution of structures. It is iteratively performed until the high-dimensional data is presented in least dimensional space. The data is visually exemplified in score plots. The relative distance among the samples in scores and loading plots depict the degree of systemic variation among metabolite samples (Boccard et al. 2010; Sugimoto et al. 2012).

- *Hierarchical Cluster Analysis (HCA)*
  Partitioning a dataset into several subclasses or clusters is the key role of clustering analysis technique. Although there are two types of cluster analysis, namely, hierarchical and nonhierarchical, both have the same function of dividing metabolomic data into subgroups of same profiles (Blekherman et al. 2011). These subgroups can unravel the inner mechanism by uncovering the underlying patterns in the data structure. A workflow of hierarchical cluster analysis can be either agglomerative or divisive, i.e., grouping objects iteratively or dividing a dataset, respectively. However, agglomerative workflow is used more often. Initially, HCA processes by calculating similarity measures among two observations followed by placing the most similar samples in same group. Once the clusters for related samples are generated, it calculates similarity among two clusters. The similarity measure is calculated using a certain metrix based on Pearson correlation, Euclidean disctance, or covariance. This procedure is performed repetitively until all the samples are aligned. A dendrogram is the graphical representation of the outcome generated, where the leaves relate to the objects and the branches portray hierarchy of clusters (Beckonert et al. 2003; Tikunov et al. 2005; Boccard et al. 2010; Sugimoto et al. 2012).

7.6.2.2 Analysis by Supervised Methods

A supervised learning method deliberates each sample with respect to observed outcome. It includes regression and classification problems based on the type of output under investigation. Multiple techniques based on machine learning or statistics have been developed for analysis under supervised conditions:

- *Partial Least Squares Regression*
  Partial least squares (PLS) regression is a linear regression-based method. It is particularly useful with a dataset that has fewer observations (samples) compared to variables (metabolites). The mechanism of PLS works on highly multidimensional data with a small number of factors, where the linear regression is then applied to these factors (Blekherman et al. 2011). PLS forms a low-

dimensional topological space based on linear combinations of X variables. Later it adjusts the model by comprehending the Y-related information in original X variables. In metabolomics, PLS-Discriminant Analysis (PLS-DA) has been most widely used due to its capability to refine the distance between different samples. PLS-DA is considered an effective tool for grouping of metabolomic data (Jonsson et al. 2005; Boccard et al. 2010). The refinement of data separation is achieved by moving the principal components to the maximum extent and then removing the variables carrying information of class separation. The graphical representation is depicted in S-plots that are similar to loading plots in PCA and allow visualization of covariance and correlation between metabolites. However, unlike PCA, it is a supervised learning technique. The S-plots contribute to the identification of potentially significant metabolites (Wiklund et al. 2008; Sugimoto et al. 2012).

- *Random Forests*
  Random Forests (RF) is categorized as a machine learning process which is usually used for depicting distinction of two datasets. As indicated by name, the "Random Forests" is basically grouping of multiple trees. The algorithm to build trees was proposed by Leo Breiman in 2001 (Breiman 2001). Each classification tree is assembled by input of sample data into the machine, and at each division the candidate set of variables is a random subset of the variables. Hence, the RF technique uses two kinds of input for tree building, i.e., bootstrap aggregation/ bagging and random variable selection. Due to bagging and random variable selection, individual trees represent low correlation (Díaz-Uriarte and De Andres 2006; Sugimoto et al. 2012). Random Forests machine learning technique has an advantage, as it is vigorous toward outliers and noise, and is simpler to interpret (Breiman 2001).

- *Artificial Neural Networks*
  Artificial neural network (ANN) is a computational model comprising of multiple units called neurons that are interlinked to form a network of connections. This technique is inspired by the structure and functions of biological neural network. The neurons in this network share similarity with biological neurons which receive an input that is a combined output of other components. Input units receive the signals which are then propagated to the output units through the network connections. When these signals arrive at the output units, they regulate activation values. Initially, random numbers are assigned to the connection weights and then gradually adjusted as the signals move toward the right output. Despite being the most popular artificial intelligence program, the output of ANNs is difficult to comprehend. Multilayer perceptron (Taylor et al. 2002) and radial basis functions (Poggio and Girosi 1990) are the most widely used algorithms of ANNs (Boccard et al. 2010). Artificial neural networks have become the most commonly used technique in MS-based studies for the identification, classification, and optimization of metabolites (Sugimoto et al. 2012).

- *Probabilistic and Instance-Based Learning Algorithms*
  Probabilistic learning algorithms are based on the probability models. A probability is assigned to each group during the training step. This algorithm works by predicting the group labels which directly correlate with the group of the highest

probability. Naïve Bayes classifier is one of the well-known methods of statistical learning among Bayesian networks. Although it is the simplest statistical learning method, many complex problems can be easily resolved by Naïve Bayes (Boccard et al. 2010).

In contrast, instance-based learning algorithms work on the principle of nearest neighbor classification. Instances with similar properties are regrouped in high-dimensional space prior to the application of k-nearest neighbor rule (Aha et al. 1991). Multiple differences among instance-based classifiers are balanced by Euclidian distances (Wettschereck et al. 1997; Boccard et al. 2010).

7.7 Analytical Techniques in Metabolomics

Metabolomics, an emerging discipline in systems biology, deals with the quantitative and qualitative analysis of a diverse class of low-molecular-weight molecules including organic compounds, vitamins, nucleic acids, lipids, carbohydrates, amino acids, and their polymers. These metabolites are either produced as an outcome of a metabolic response or synthesized in a pathophysiological condition. Owing to the diverse properties and complexity of metabolites, a single instrumental technique cannot completely analyze the whole plant metabolome. Therefore, separation techniques such as chromatography and electrophoresis integrated with detection techniques, i.e., mass, NMR, and vibrational spectroscopy, are applied, which enable quick separation, efficient isolation, reliable detection, systematic characterization, and precise quantification of metabolites in a bio-fluid sample. The following are the details of few techniques which are frequently used in metabolome analysis:

7.7.1 Mass Spectrometry-Based Methods (MS)

Mass spectroscopy (MS) is one of the most frequently used techniques in metabolomics. Coupled with other separation techniques, this analytical method is very sensitive and covers a wide range of metabolites. The MS technique follows three fundamental steps, namely, (1) ion formation, (2) ion separation, and (3) ion detection. The formation of ions can be carried out by either hard or soft ionization sources. The hard ionization process involves electron bombardment to the gaseous or vaporous state of analytes, also known as electron ionization-mass spectrometry (EI-MS) technique, invoking a high degree of fragmentation. Although this technique offers a detailed mass spectrum which, following careful analysis, can provide important information regarding structure elucidation of unknown metabolite, it is generally not coupled with HPLC owing to different characteristics of HPLC and EI-MS, where the former technique works under high atmospheric pressure for efficient separation, while in the later technique, the ionization is fundamentally accomplished under high vacuum. The EI technique is, therefore, mostly used in GC-MS techniques where the entire system is under high vacuum. Contrary to hard ionization, soft ionization sources produce fewer fragmentations and, therefore, can conclusively yield the molecular mass peak of the compound. Soft ionization is achieved through a wide range of techniques, such as (1) electrospray ionization (ESI), where a high voltage (i.e., strong electric field) is applied to a liquid analyte passing through a capillary tube in order to create fine aerosol; (2) chemical ionization (CI), in which the ions of some reagent gas such as methane or ammonia present at the source strike with analyte to produce fragments; (3) fast-atom bombardment (FAB) and liquid secondary ion mass spectrometry (LSIMS) methods that use a beam of high-energy Xe or Ar atoms and Cs^+ ions under vacuum, respectively, that strike with the analytes dissolved in a liquid matrix such as glycerol or m-nitrobenzyl alcohol (NBA) and generates ions; and (4) matrix-assisted laser desorption/ionization (MALDI), unlike FAB/LSIMS, which uses a laser beam that is fired at the crystalline matrix in the dried-droplet spot-containing analyte to produce ions. Various kinds of laser sources can be used such as UV laser (nitrogen lasers), neodymium-doped yttrium aluminum garnet, IR laser (such as erbium-doped yttrium aluminum garnet), etc. The matrix crystals are generally made up of 3,5-dimethoxy-4-hydroxycinnamic acid (sinapinic acid), α-cyano-4-hydroxycinnamic acid (α-CHCA, alpha-cyano, or alpha-matrix), or 2,5-dihydroxybenzoic acid (DHB or gentisic acid). As soon as the ions are formed, they are accelerated through the mass analyzer where the ions are separated according to their mass-to-charge (m/z) ratio, a unitless entity. The acceleration of ions is achieved by electrical and/or magnetic field(s) which affect ion trajectory and velocity. The separation of ions is governed by Lorentz force law and Newton's second law of motion. The mass analyzer can be of six general types:

Mass analyzer	Ion acceleration method	Principle of ion separation
Magnetic sector mass analyzer	Magnetic fields	The applied magnetic field bends the ion beam in an arc with a unique radius. The separation of ions is based on its momentum, i.e., ions with greater momentum will follow an arc with a larger radius. In other words, ions with constant kinetic energy but different mass are separated by their trajectories in magnetic field
Electrostatic sector mass analyzer	Electric fields	As the ions travel through the electric field, they are deflected and follow the curve of the analyzer. The radius of the trajectory of ions depends on the kinetic energy of the ions and the potential field applied across the plates. Generally, the electrostatic sector analyzer is not useful alone. When electrostatic and magnetic sector analyzers are separately employed as a stand-alone unit in MS, they are termed as single-focusing instruments. When both analyzers are combined in MS, they are called a double-focusing instrument, as the instrument focuses on both the energies and the angular dispersions

Mass analyzer	Ion acceleration method	Principle of ion separation
Time-of-flight (TOF) mass analyzer	Electric fields	In TOF, separation is based on the kinetic energy and velocity of the ions. Initially all ions are accelerated in an electric field, and as a result identically charged ions have the same kinetic energy. However, in the field free flight tube ions acquire different velocities depending on their masses. The lighter mass ions get faster and, therefore, reach the detector in a shorter period of time
Quadrupole mass analyzer	Electric fields	In quadrupole, the oscillating electric fields along with static direct current are applied that lead to the separation of ions based on the stability of their trajectories. In high radio-frequency (RF) voltage, the trajectories of light mass ions become unstable and collide with the wall of rod, while the heavier mass ions produce stable trajectories and reach the detector
Quadrupole ion trap mass analyzer	Electric fields	It works on the same principle as quadrupole mass analyzer except that the analyzer is made with a doughnut-shaped ring electrode and two grounded endcap electrodes instead of metal rods. Ions are trapped for a fixed period of time in a central chamber, and later they are sequentially released based on their m/z values and reach the detector.
Ion cyclotron resonance (ICR) mass analyzer	Magnetic fields	In ICR, the mass-to-charge ratio (m/z) of ions is determined based on the cyclotron frequency of ions in a magnetic field. Unlike quadrupole ion trap mass analyzer, the ions in ICR are trapped in a penning trap under magnetic field. The ions are then excited in a larger cyclotron radius by an oscillating electric field. As the excitation field is removed, the ions rotate at their cyclotron frequency and induce a charge, which is detected as an image current that results in a signal called free induction decay (FID)

These separated ions enter the detector, which records either the ionic charge or the current produced when an ion passes by or hits a surface. There are various types of detectors which are used in mass spectroscopy. Faraday cup collector – a simple metallic conductive cup attached to a circuit - is perhaps the oldest detector, which measures the current produced when an ion strikes the cup in a vacuum and becomes neutralized. Electron multipliers are among the most commonly used detectors in modern mass spectrometry, where the ion beam strikes the dynode plate and initiates secondary electron emission process. The electrons emitted hit the next dynode surface to induce the next process of secondary emission, leading to more electrons. The amplification process is repeated until a large number of electrons are collected by a metal anode, and it can generate a measurable voltage pulse. Another commonly used detector in MS is called photomultiplier or scintillation counter, where the ions first strike a dynode and induce the secondary electron emission process. These emitted electrons then impinge upon a thin disk of crystalline phosphor and thus produce scintillation. The resultant photons enter a photomultiplier tube and impact on photocathode, which ultimately produces photoelectrons through a phenomenon known as photoelectric effect followed by electronic amplification and detection as achieved in electron multipliers.

- *Gas chromatography-mass spectrometry (GC-MS)*
  GC-MS is among the widely used analytical techniques in metabolomics that integrates the features of gas chromatography with mass spectroscopy. The method is highly sensitive, reproducible, and well established, and it can detect and identify minute quantity of small organic compounds from a test sample. This method, therefore, finds a wide range of applications in metabolomic profiling, medicine, astrochemistry, chemical engineering, food and beverages, drug and narcotic detection, environment analysis, criminal forensics, explosive, chemical warfare detection, etc. The method uses the gas chromatography technique for compound separation; only those compounds which are volatile or can be made volatile through derivatization are detected by MS. The fact that plant metabolites contain a number of functional groups, the derivatization is carried out in two steps. In the first step, the carbonyl functional groups that are present in metabolites are protected by converting them into oximes with O-alkylhydroxylamine followed by the reaction with silylating reagents such as N-methyl-N-(trimethylsilyl) trifluoroacetamide, etc. that replace exchangeable protons and form trimethylsilyl (TMS) esters. Metabolites such as amino acids and carbohydrates contain a number of exchangeable protons, and therefore, a range of derivatized products are formed, whereas organic acids often yield one product. GC-MS has been successfully used to study the effects of genetic or environmental modifications and stressors in several plants including *Arabidopsis*, potatoes, and tomatoes by analyzing either intracellular or volatile metabolites (Lisec et al. 2006; Roessner et al. 2000; Fiehn et al. 2000; Roessner-Tunali et al. 2003).
- *Liquid chromatography-mass spectrometry (LC-MS)*
  LC-MS is another powerful and widely employed analytical technique used to separate, detect, and quantify metabolites that may or may not be amenable to GC-MS analysis, thereby increasing the applicability of this technique in metabolomics. The method combines the separation capability of liquid chromatography such as HPLC with MS. The ultra-performance liquid chromatography (UPLC) can also be used instead of HPLC to further improve the separation efficiency and peak resolution.

  The LC uses a highly pressurized liquid mobile phase, while the mass analyzer in spectrometer operates under high vacuum; both devices are fundamentally incompatible unlike GC-MS, and therefore an interface is required to transfer the separated compounds from LC device to mass spectrometer source smoothly. The LC-MS interface removes the mobile phase from elute while preserving the analyte by using strategies of atmospheric pressure ionization (API) techniques such as electrospray ionization (ESI), atmospheric pressure chemical ionization (APCI), and atmospheric pressure photoionization (APPI). Several plant-based metabolomics studies have been performed using LC-MS techniques (De Vos et al. 2007; van der Hooft et al. 2012). Moco and co-workers have also developed a Metabolome Tomato Database (MoTo DB) that is dedicated to LC-MS-based metabolomics of tomato fruit (Moco et al. 2006).

- *Capillary electrophoresis-mass spectrometry (CE-MS)*

  CE-MS is another powerful analytical technique in metabolomics that permits the separation of charged metabolites in liquid electrolytic solution under the influence of a spatially uniform electric field. Metabolites are separated based on their electrophoretic mobility, where their size, viscosity, and charge play a key role. CE is generally coupled with electrospray mass spectrometry; however, it could also be combined with FAB, MALDI, and APCI. The CE-MS finds broad applications in proteomics, environmental science, forensics, and clinical medicine as well as in quantitative analysis of biomolecules.

7.7.2 Nuclear Magnetic Resonance (NMR) Spectroscopy

NMR spectroscopy is an extremely powerful technique in metabolomics and natural product chemistry, which is widely used in the elucidation of metabolite structure. Unlike mass spectrometry, NMR is a nondestructive technique and is based on the magnetic properties of the atomic nuclei which can provide significant information about a metabolite. However, only few atoms with either odd atomic number or odd mass number such as ^{1}H, ^{13}C, ^{15}N, ^{19}F, and ^{31}P are NMR active, as they possess nuclear spin in the presence of a magnetic field. Depending on the chemical environment of these atoms, their nuclei absorb energy at different radio frequencies (RF) causing the nuclei to be promoted from low-energy spin states to high-energy spin states, something similar to electronic excitation. The subsequent release of energy over a period of time, known as relaxation process, is measured in the form of free indication decay (FID). A Fourier transformation is then applied to convert the time domain FID into a frequency domain dataset, which leads to the construction of NMR chemical shift (in ppm) against the intensity of peak. The 1H-NMR is the most frequently used technique in metabolomics, where the chemical shift values of sample molecule are normalized against a reference D_2O solution of tetramethylsilane (TMS), which is set to as 0 ppm. The typical range of measured chemical shift for ^{1}H-NMR is 0–10 ppm, and for ^{13}C it varies from 0 to 250 ppm. The intensity of the peak depends on the number of identical nuclei and thus allows the quantification of an atom. Typically, multiple peaks at different chemical shift values are observed for a single metabolite; for instance, a glucose molecule may have roughly 30 peaks referring to ^{1}H and ^{13}C atoms at different positions. Generally, for the structure elucidation of a metabolite, different NMR experiments are required, such as one-dimensional and two-dimensional ^{1}H-NMR, ^{13}C NMR, as well as standard ^{1}H, J-resolved, ^{1}H-^{1}H correlation spectroscopy (COSY), ^{1}H-^{1}H total correlation spectroscopy (TOCSY), ^{1}H-^{13}C heteronuclear single-quantum correlation (HSQC), heteronuclear multiple bond correlation (HMBC), etc. NMR has been used for many years in the analysis of primary (e.g., sugars and amino acids) and secondary metabolites such as phenolic compounds of plant (Kim et al. 2010). Likewise, the metabolic profile of the seeds of seven chia (*Salvia hispanica* L.) populations has been also successfully investigated by NMR spectroscopy integrated with principal component analysis and chemometrics (de Falco et al. 2017). Also, NMR has been also used in the analysis of various plant extracts such as tobacco and *Arabidopsis* (Ward et al. 2003; Kikuchi et al. 2004; Choi et al. 2004). Sekiyama et al. have recently reported the NMR-based metabolic profiling of field-grown leaves for 12 genotypes of sugar beet (*Beta vulgaris* L.) exhibiting different levels of *Cercospora* leaf spot (CLS) resistance (Sekiyama et al. 2017).

Recently, researchers have also worked on combining two powerful analytical approaches in metabolomics such as NMR and MS (Bingol and Brüschweiler 2015; Farag et al. 2012) to improve metabolomics sample characterization in terms of identification and quantification, as well as in accelerating new metabolite discovery.

7.7.3 Vibrational Spectroscopy

When compounds are exposed to an electromagnetic (EM) radiation such as UV or IR, they absorb light of different wavelengths. The absorbed light causes the chemical bond to vibrate in various ways such as stretching and bending, etc. The fact that one specific bond or angle vibrates in a specific way by absorbing a specific wavelength of EM can help in the identification of functional group present in the metabolites. For instance, in fatty acid molecules, the stretching vibrations of C-H bond in -CH_3 and -CH_2- groups can range from 2800 to 3050 cm^{-1} in IR spectra. Likewise, the region between 1500 and 1800 cm^{-1} is dominated by amide I (>C=O) and amide II (>C-NH) stretching, indicating the predominance of either alpha-helix or beta-sheet structures. Similarly a sharp band around 2200–2400 cm^{-1} indicates the presence of -C≡N or -C≡C- bonds. A typical absorption region for covalent bonds ranges from 600 to 4000 cm^{-1}. Due to its holistic nature and the ability to identify functional groups, vibrational spectroscopy, such as Fourier transform infrared (FT-IR) and Raman spectroscopy, is a valuable, nondestructive, analytical technique for metabolic profiling and can analyze a range of primary and secondary metabolites including amino acids, carbohydrates, lipids, and fatty acids as well as many other organic compounds.

7.8 Metabolomics: The Apogee in Realm of OMICS

Over the past decades, with the advent of OMICS techniques (such as genomics, proteomics, and metabolomics), the concept of system biology and integrated OMICS is gaining attention for unraveling complex processes at molecular and organismal levels (Wanichthanarak et al. 2015).

Metabolomics is downstream of transcriptomics and proteomics. Unlike two others, the size of metabolome of a species cannot be hypothesized by tools that use existing genomic information on central dogma principle. Analysis of intricate metabolite interactions for key players of pathways leads to significant understanding of individual genomic information and metabolic outputs (Toubiana et al. 2013). Similarly, while moving toward the concept of system biology, metabolites, proteins, and genes data should be analyzed in an integrated way to build the complete picture of pathways for plant regulatory responses under question.

Low-cost, efficient, and high-yielding approaches and software tools are magically increasing omics data by adding new plant species and sequence information (Kleessen and Nikoloski 2012; Rai and Saito 2016). Various approaches are in use to integrate multiple omic datasets such as:

- Pathway- or Biochemical-Ontology-Based Integration
  Examples of software/tools: Impala, iPEAP
- Biological-Network-Based Integration
  Examples of software/tools: SAMNetWeb, pwOmics, MetaMapR, MetScape
- Empirical Correlation Analysis
  Examples of software/tools: WGCNA, MixOmic, DiffCorr, qpgraph

Metabolites are direct impressions of biochemical activity and hence have higher correlation with phenotype than genes and proteins (Tohge et al. 2014). It is a well-known fact that phenotypic diversity is determined by genetic as well as environmental contributions. To find out underlying associations among genetic variation (mostly qualitative) and phenotypic variation (mostly quantitative), various approaches have been proposed. This association laid the foundation for quantitative trait locus (QTL) mapping and genome-wide association studies (GWAS). Such analyses are most widely used to explore phenotypic impact, number, and interaction of genes responsible for targeted quantitative trait (Wu et al. 2016)

Microarrays offer efficient data for small populations as they are expensive and hence not suitable for complex physiological traits where large populations are required to be analyzed. Moreover, traits with moderate to low heritability are not dissectible as only those with higher heritability are easy to be phenotyped with microarrays. On the contrary, higher-throughput and low-cost metabolomics platforms made these an ideal choice for genetic and genomic studies. Choice of experiment design, population, variables for mapping, and platform is very important for target achievement (Weckwerth 2003; Wen et al. 2018)

Metabolites have been phenotyped in various plant species, to unveil related genes. Metabolic QTL (mQTL) and metabolome-genome-wide association studies (mGWAS) started on model plants and extended to crop species and gained attention with advent of high-throughput metabolic platforms and sequencing technologies. The number of studies with mQTLs and mGWAS is increasing every day. MS-based primary metabolites were profiled using *Arabidopsis* recombinant inbred line (RIL) populations (Col-0 × Col-24) which resulted in important genetic signatures. This study provided the basis for identification of hitherto valuable unknown genes for improving seed metabolism (Knoch et al. 2017). The Arabidopsis 1001 genome project, which started in 2008, presents a rich source of information for scientists. Sequence data offers a huge opportunity of performing GWAS in such big collections of natural populations. Targeted (Francisco et al. 2016) and nontargeted (Wu et al. 2018) metabolite-based GWAS have identified various important genetic loci. In addition, abiotic stress-related metabolite profiling in control and stressed plants using 309 *Arabidopsis* accessions resulted in 70 putative associations among metabolites and structural genes. Eight of the associations were experimentally validated using tDNA knockouts, out of which two (related to purine nucleotide metabolism and lysine degradation) were found with differential expression under two different experimental conditions (Wu et al. 2018).

Cereals are the most important dietary components and are a direct or indirect focus of quality and quantity improvement research. Genes related to variation in rice secondary metabolism have been identified using mGWAS. Metabolite data for 175 rice accessions revealed that one third of metabolites are controlled by major mQTLs, while for intraspecies diversity of metabolites, few loci were detected. This study further highlighted the diversity of phytochemicals in this plant species (Matsuda et al. 2015). In wheat, 197 metabolic features in flag leaves were identified using 179 doubled haploid lines. This data together with mapping of agronomic traits helped in understanding the genetic basis of variations, covariations, and correlations among agronomic traits, metabolic signatures, and plant phenology (Hill et al. 2015). Moreover, in maize mGWAS were performed on various tissues such as kernel, leaf, and seed using various methods such as UPLC, GC-MS, and HPLC (Riedelsheimer et al. 2012; Lipka et al. 2013; Luo 2015; Ramalingam et al. 2015). Wen et al. reported GWAS based on metabolome profiling of primary metabolites in four different tissues: seedling leaf, leaf at reproductive stage, young kernel, and mature kernel. 38 of 61 identified metabolites were found common among abovementioned tissues. Many known alongwith less recognized genes were identified that were tissue specific as well as common among all tissues (Wen et al. 2018). Fruit primary metabolites were studied in tomato using 163 accessions. Forty-four associations were identified in 36 traits using multi-locus mixed model GWAS (Sauvage et al. 2014).

In addition to experimental integrated studies, various databases for inter-species analysis, integrated as well as individual OMICS levels has also been developed for both model and crop species (Mutwil et al. 2011; Hamada et al. 2011; Sato et al. 2013; Obayashi et al. 2014). The Plant Omics Data Center was developed as gene expression network (GEN) repository, initially for eight plant species and now updated for ten species with addition of *Zea mays*, *Nicotiana tabacum*, and *Physcomitrella patens* (Ohyanagi et al. 2015). Databases for individual species with multi-omics information such as TAIR for *Arabidopsis* (Lamesch et al. 2012), TOMATOMICS for tomato (Kudo et al. 2017), and UniVIO for rice (Kudo et al. 2013) are also contributing significantly for system biology research.

7.9 Challenges and Perspective in Agricultural Metabolomics

Metabolomics is characterized as an effective strategy for distinguishing proof and evaluation of low-atomic weight metabolites in an organic sample (Saito and Matsuda 2010). This approach is increasingly applied in many crop species in the last two decades, regardless of transgenic system availability (Fernie and Schauer 2009; Simó et al. 2014; kumar et al. 2017).

It is a powerful omics tool for selection of beneficial traits, eventually for improvement of crop species (Tian et al. 2016; Zivy et al. 2015). In addition, metabolomics also facilitates study of systems biology, particularly by providing important clues in understanding interactions of plants and their interacting environments as well as humans and plants by unraveling the medicinal plants' beneficial compounds (Shyur and Ynag 2008; Taketo et al. 2010).

Since plant metabolomics is an emerging field, many challenges are still faced by researchers, and consequently various aspects in analytical, systematics, and computational level need to be addressed. These include the fact that data obtained after metabolite profiling is itself not sufficient to build pathways or to determine regulatory elements. This calls for integrated omics studies (Fig. 7.4); however this task is even more challenging due to the complexity of omics relationships (Zierer et al. 2015; Bersanelli et al. 2016; Huang et al. 2017).

Metabolites are influenced by many external as well as internal factors that are faced by a plant during its life. They are beneficial for providing opportunities to

Fig. 7.4 Workflow for OMICS data integration. Only metabolomics interactions are depicted

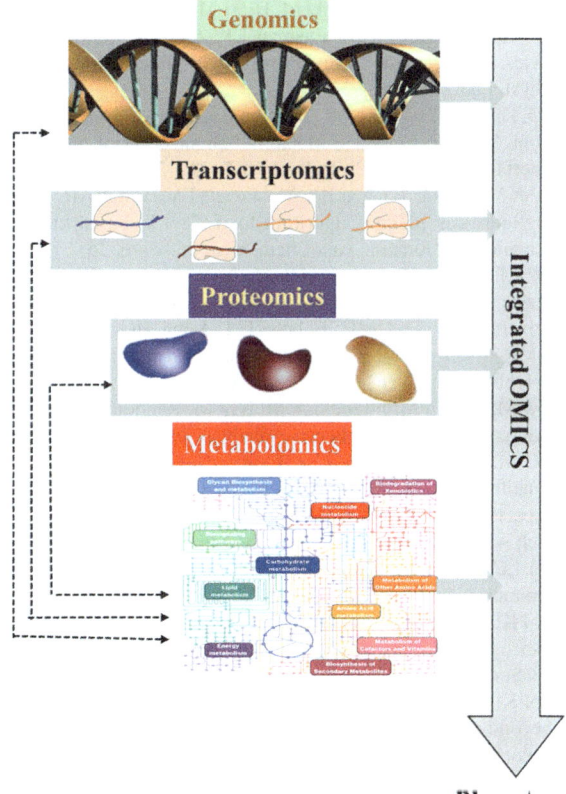

acclimatize the plants to extreme environmental conditions (Fernie and Klee 2011). This could be done via metabolite engineering, but it would be difficult to alter expression of all controlling metabolites. Metabolomics could also be used as taxonomic tool for plant phylogenetic analysis too. In addition, some new species-metabolites relations have also been reported in result of this analysis (Liu et al. 2017).

Taking the challenges detailed above into account, future strategies are needed with a strong focus on computational metabolomics tools for high-resolution and broader metabolome coverage (Johnson et al. 2015). Development of metabolomic platforms is required for efficient and accurate qualitative as well as quantitative identification of as many as possible molecules (metabolites). Bioinformatics tools, as a backbone for analysis and integration of data, will also help to increase precision of data interpretation and development/correction of metabolic pathways as well as more systematic plant phylogenetic analyses.

References

Afendi FM, Okada T, Yamazaki M, Hirai-Morita A, Nakamura Y, Nakamura K, Ikeda S, Takahashi H, Altaf-Ul-Amin M, Darusman LK, Saito K, Kanaya S (2012) KNApSAcK family databases: integrated metabolite–plant species databases for multifaceted plant research. Plant Cell Physiol 53(2):e1–e1

Aha D, Kibler WD, Albert MK (1991) Instance-based learning algorithms. Mach Learn 6(1):37–66

Aharoni A, Galili G (2011) Metabolic engineering of the plant primary–secondary metabolism interface. Curr Opin Biotechnol 22(2):239–244

Andreev VP, Rejtar T, Chen HS, Moskovets EV, Ivanov AR, Karger BL (2003) A universal denoising and peak picking algorithm for LC– MS based on matched filtration in the chromatographic time domain. Anal Chem 75(22):6314–6326

Bais P, Moon SM, He K, Leitao R, Dreher K, Walk T, Wurtele ES (2010) PlantMetabolomics.org: a web portal for plant metabolomics experiments. Plant Physiol 152(4):1807–1816

Beckonert O, Bollard ME, Ebbels TM, Keun HC, Antti H, Holmes E, Nicholson JK (2003) NMR-based metabonomic toxicity classification: hierarchical cluster analysis and k-nearest-neighbour approaches. Anal Chim Acta 490(1–2):3–15

Bellew M, Coram M, Fitzgibbon M, Igra M, Randolph T, Wang P, Chen J (2006) A suite of algorithms for the comprehensive analysis of complex protein mixtures using high-resolution LC-MS. Bioinformatics 22(15):1902–1909

Bersanelli M, Mosca E et al (2016) Methods for the integration of multi-omics data: mathematical aspects. BMC Bioinforma 17(Suppl 2):15

Bertsch A, Gröpl C, Reinert K, Kohlbacher O (2011) Open MS and TOPP: open source software for LC-MS data analysis. In M. Hamacher, M. Eisenacher, & C. Stephan.Data mining in proteomics. New York, NY: Humana Press, p 353–367

Bingol K, Brüschweiler R (2015) Two elephants in the room: new hybrid nuclear magnetic resonance and mass spectrometry approaches for metabolomics. Curr Opin Clin Nutr Metab Care 18(5):471–477

Bingol K, Bruschweiler-Li L, Yu C, Somogyi A, Zhang F, Brüschweiler R (2015) Metabolomics beyond spectroscopic databases: a combined MS/NMR strategy for the rapid identification of new metabolites in complex mixtures. Anal Chem 87(7):3864–3870

Blekherman G, Laubenbacher R, Cortes DF, Mendes P, Torti FM, Akman S, Shulaev V (2011) Bioinformatics tools for cancer metabolomics. Metabolomics 7(3):329–343

Boccard J, Veuthey JL, Rudaz S (2010) Knowledge discovery in metabolomics: an overview of MS data handling. J Sep Sci 33(3):290–304

Breiman L (2001) Random forests. Mach Learn 45(1):5–32

Brown DE, Rashotte AM, Murphy AS, Normanly J, Tague BW, Peer WA, Taiz L, Muday GK (2001) Flavonoids act as negative regulators of auxin transport in vivo in Arabidopsis. Plant Physiol 126:524–535

Bundy JG, Davey MP, Viant MR (2008) Environmental metabolomics: a critical review and future perspectives. Metabolomics 5(1):3–21

Caretto S, Linsalata V, Colella G, Giovanni M, Lattanzio V (2015) Carbon fluxes between primary metabolism and phenolic pathway in plant tissues under stress. Int J Mol Sci 16:26378–26394

Castillo-Ordóñez WO, Tamarozzi ER et al (2017) Exploration of the acetylcholinesterase inhibitory activity of some alkaloids from Amaryllidaceae family by molecular docking in silico. Neurochem Res 42(10):2826–2830

Chandra H, Bishnoi P et al (2017) Antimicrobial resistance and the alternative resources with special emphasis on plant-based antimicrobials—a review. Plan Theory 6(2):16

Choi HK, Choi YH, Verberne M, Lefeber AW, Erkelens C, Verpoorte R (2004) Metabolic fingerprinting of wild type and transgenic tobacco plants by 1H NMR and multivariate analysis technique. Phytochemistry 65(7):857–864

Conforti F, Loizzo MR et al (2010) Quantitative determination of Amaryllidaceae alkaloids from Galanthus reginae-olgae subsp. vernalis and in vitro activities relevant for neurodegenerative diseases. Pharm Biol 48(1):2–9

Cragg GM, Newman DJ (2013) Natural products: a continuing source of novel drug leads. Biochim Biophys Acta Gen Subj 1830(6):3670–3695

Cui Q, Lewis IA, Hegeman AD, Anderson ME, Li J, Schulte CF, Markley JL (2008) Metabolite identification via the madison metabolomics consortium database. Nat Biotechnol 26(2):162

Daskalchuk T, Ahiahonu P, Heath D, Yamazaki Y (2006) The use of non-targeted metabolomics in plant science. In: Plant metabolomics. Springer, Berlin, pp 311–325

Daviss, Bennett (2005) Growing pains for metabolomics. Scientist 19(8):25–28

de Falco BG, Incerti et al (2017) Metabolomic analysis of *Salvia hispanica* seeds using NMR spectroscopy and multivariate data analysis. Ind Crop Prod 99:86–96

De Souza DP, Saunders EC, McConville MJ, Likić VA (2006) Progressive peak clustering in GC-MS Metabolomic experiments applied to Leishmania parasites. Bioinformatics 22(11):1391–1396

De Vos RC, Moco S, Lommen A, Keurentjes JJ, Bino RJ, Hall RD (2007) Untargeted large-scale plant metabolomics using liquid chromatography coupled to mass spectrometry. Nat Protoc 2(4):778–791

Defernez M, Le Gall G (2013) Strategies for data handling and statistical analysis in metabolomics studies. In In: Rolin D (ed) Advances in botanical research, vol 67. Elsevier Academic Press, San Diego, USA, pp 493–555

Dettmer K, Aronov PA et al (2007) Mass spectrometry-based metabolomics. Mass Spectrom Rev 26(1):51–78

Díaz-Uriarte R, De Andres SA (2006) Gene selection and classification of microarray data using random forest. BMC Bioinforma 7(1):3

Edward HD (2014) Biochemical facts behind the definition and properties of metabolites, Ebook, Biochemistry and Biophysics and Faculty of Nutrition Texas A&M University

Farag MA, Porzel A, Schmidt J, Wessjohann LA (2012) Metabolite profiling and fingerprinting of commercial cultivars of *Humulus lupulus L.* (hop): a comparison of MS and NMR methods in metabolomics. Metabolomics 8(3):492–507

Fernie AR, Klee HJ (2011) The use of natural genetic diversity in the understanding of metabolic organization and regulation. Front Plant Sci 2:59

Fernie AR, Schauer N (2009) Metabolomics-assisted breeding: a viable option for crop improvement? Trends Genet 25(1):39–48

Fiehn O (2001) Combining genomics, metabolome analysis, and biochemical modelling to understand metabolic networks. Comp Funct Genomics 2(3):155–168

Fiehn O (2002) Metabolomics-the link between genotypes and phenotypes. Plant Mol Biol 48:155–171

Fiehn O, Kopka J, Dörmann P, Altmann T, Trethewey RN, Willmitzer L (2000) Metabolite profiling for plant functional genomics. Nat Biotechnol 18(11):1157–1161

Francisco M, Joseph B et al (2016) Genome wide association mapping in *Arabidopsis thaliana* identifies novel genes involved in linking allyl glucosinolate to altered biomass and defense. Front Plant Sci 7:1010

Fukushima A, Kusano M (2013) Recent progress in the development of metabolome databases for plant systems biology. Front Plant Sci 4:73

George P (2005) Metabolomics comes of age? Scientist 19(11):8

Gupta OP, Karkute SG, Banerjee S, Meena NL, Dahuja A (2017) Contemporary understanding of miRNA-based regulation of secondary metabolites biosynthesis in plants. Front Plant Sci 8:374. https://doi.org/10.3389/fpls.2017.00374

Hamada K, Hongo K et al (2011) OryzaExpress: an integrated database of gene expression networks and omics annotations in rice. Plant Cell Physiol 52(2):220–229

Hartmann T, Kutchan TM, Strack D (2005) Evolution of metabolic diversity. Phytochemistry 66:1198–1199

Hegeman AD (2010) Plant metabolomics—meeting the analytical challenges of comprehensive metabolite analysis. Brief Funct Genomics 9(2):139–148

Herwig OG, Ludwig-Muller J (2014) Plant natural products: synthesis, biological functions and practical applications, 1st edn. Wiley-VCH Verlag GmbH & Co. KGaA, Weinheim

Hill CB, Taylor JD et al (2015) Detection of QTL for metabolic and agronomic traits in wheat with adjustments for variation at genetic loci that affect plant phenology. Plant Sci 233:143–154

Horai H, Arita M, Kanaya S, Nihei Y, Ikeda T, Suwa K, Ojima Y, Tanaka K, Tanaka S, Aoshima K et al (2010) MassBank: a public repository for sharing mass spectral data for life sciences. J Mass Spectrom 45:703–714

Huang S, Chaudhary K et al (2017) More is better: recent progress in multi-omics data integration methods. Front Genet 8:84

Iijima Y, Nakamura Y, Ogata Y, Tanaka K, Sakurai N, Suda K, Suzuki T, Suzuki H, Okazaki K, Kitayama M, Kanaya S, Aoki K, Shibata D (2008) Metabolite annotations based on the integration of mass spectral information. Plant 54(5):949–962

Johnson SR, Lange BM (2015) Open-access metabolomics databases for natural product research: present capabilities and future potential. Front Bioeng Biotechnol 3:22

Johnson CH, Ivanisevic J et al (2015) Bioinformatics: the next frontier of metabolomics. Anal Chem 87(1):147–156

Jonsson P, Bruce SJ, Moritz T, Trygg J, Sjöström M, Plumb R, Antti H (2005) Extraction, interpretation and validation of information for comparing samples in metabolic LC/MS data sets. Analyst 130(5):701–707

Jordan KW, Nordenstam J, Lauwers GY, Rothenberger DA, Alavi K, Garwood M, Cheng LL (2009) Metabolomic characterization of human rectal adenocarcinoma with intact tissue magnetic resonance spectroscopy. Dis Colon Rectum 52(3):520–525

Jorge TF, Rodrigues JA et al (2016) Mass spectrometry-based plant metabolomics: metabolite responses to abiotic stress. Mass Spectrom Rev 35(5):620–649

Kanehisa M, Goto S (2000) KEGG: Kyoto encyclopedia of genes and genomes. Nucleic Acids Res 28(1):27–30

Katajamaa M, Orešič M (2007) Data processing for mass spectrometry-based metabolomics. J Chromatogr A 1158(1–2):318–328

Katajamaa M, Miettinen J, Orešič M (2006) MZmine: toolbox for processing and visualization of mass spectrometry based molecular profile data. Bioinformatics 22(5):634–636

Kikuchi J, Shinozaki K, Hirayama T (2004) Stable isotope labeling of *Arabidopsis thaliana* for an NMR-based metabolomics approach. Plant Cell Physiol 45(8):1099–1104

Kim HK, Choi YH, Verpoorte R (2010) NMR-based metabolomic analysis of plants. Nat Protoc 5(3):536–549

Kleessen S, Nikoloski Z (2012) Dynamic regulatory on/off minimization for biological systems under internal temporal perturbations. BMC Syst Biol 6:16–16

Knoch D, Riewe D et al (2017) Genetic dissection of metabolite variation in Arabidopsis seeds: evidence for mQTL hotspots and a master regulatory locus of seed metabolism. J Exp Bot 68(7):1655–1667

Krishnan P, Kruger NJ et al (2005) Metabolite fingerprinting and profiling in plants using NMR. J Exp Bot 56(410):255–265

Kudo T, Akiyama K et al (2013) UniVIO: a multiple omics database with hormonome and transcriptome data from rice. Plant Cell Physiol 54(2):e9–e9

Kudo T, Kobayashi M et al (2017) TOMATOMICS: a web database for integrated omics information in tomato. Plant Cell Physiol 58(1):e8–e8

Kumar A, Yadav A, Gupta N, Kumar S, Gupta N, Kumar S, Gurjar H (2015) Metabolites in plants and its classification. World J Pharm Pharm 4:287–305

Kumar R, Bohra A et al (2017) Metabolomics for plant improvement: status and prospects. Front Plant Sci 8:1302

Lamesch P, Berardini TZ et al (2012) The Arabidopsis Information Resource (TAIR): improved gene annotation and new tools. Nucleic Acids Res 40(Database issue):D1202–D1210

Last RL, Jones AD et al (2007) Towards the plant metabolome and beyond. Nat Rev Mol Cell Biol 8:167

Lipka AE, Gore MA et al (2013) Genome-wide association study and pathway-level analysis of tocochromanol levels in maize grain. G3: Genes Genomes Genet 3(8):1287–1299

Lisec J, Schauer N, Kopka J, Willmitzer L, Fernie AR (2006) Gas chromatography mass spectrometry-based metabolite profiling in plants. Nat Protoc 1(1):387–396

Liu K, Abdullah AA et al (2017) Novel approach to classify plants based on metabolite-content similarity. Biomed Res Int 2017:5296729

Lundberg P, Vogel T, Malusek A, Lundquist PO, Cohen L, Dahlqvist O (2005) MDL – the magnetic resonance metabolomics database (mdl.imv.liu.se). ESMRMB, Basel

Luo J (2015) Metabolite-based genome-wide association studies in plants. Curr Opin Plant Biol 24:31–38

Mahdavi V, Farimani MM et al (2015) A targeted metabolomics approach toward understanding metabolic variations in rice under pesticide stress. Anal Biochem 478:65–72

Malitsky S, Blum E, Less H, Venger I, Elbaz M, Morin S, Eshed Y, Aharoni A (2008) The transcript and metabolite networks affected by the two clades of Arabidopsis glucosinolate biosynthesis regulators. Plant Physiol 148:2021–2049

Matsuda F, Nakabayashi R et al (2015) Metabolome-genome-wide association study dissects genetic architecture for generating natural variation in rice secondary metabolism. Plant J 81(1):13–23

Meier R, Ruttkies C et al (2017) Bioinformatics can boost metabolomics research. J Biotechnol 261:137–141

Mirza SB, Bokhari H, Fatmi MQ (2015) Exploring natural products from the biodiversity of Pakistan for computational drug discovery studies: collection, optimization, design and development of a chemical database (ChemDP). Curr Comput Aided Drug Des 11(2):102–109

Moco S, Bino RJ, Vorst O, Verhoeven HA, de Groot J, van Beek TA, Vervoort J, de Vos CH (2006) A liquid chromatography-mass spectrometry-based metabolome database for tomato. Plant Physiol 141(4):1205–1218

Mutwil M, Klie S et al (2011) PlaNet: combined sequence and expression comparisons across plant networks derived from seven species. Plant Cell 23(3):895–910

Ncube B, Staden J (2015) Tilting plant metabolism for improved metabolite biosynthesis and enhanced human benefit. Molecules 20:12698–12731

Ncube B, Finnie J, Van Staden J (2012) Quality from the field: the impact of environmental factors as quality determinants in medicinal plants. S Afr J Bot 82:11–20

Nielsen J, Oliver S (2005) The next wave in metabolome analysis. Trends Biotechnol 23(11): 544–546

Nielsen NPV, Carstensen JM, Smedsgaard J (1998) Aligning of single and multiple wavelength chromatographic profiles for chemometric data analysis using correlation optimised warping. J Chromatogr A 805(1–2):17–35

Nix HA (1990) National geographic information system – an achievable objective? In: Keynote address, Aurisa

Noel JP, Austin MB, Bomati EK (2005) Structure–function relationships in plant phenylpropanoid biosynthesis. Curr Opin Plant Biol 8:249–253

Obayashi T, Okamura Y et al (2014) Evaluation of gene coexpression in agriculturally important plants. Plant Cell Physiol 55(1):e6–e6

Ohyanagi H, Takano T et al (2015) Plant omics data center: an integrated web repository for inter-species gene expression networks with NLP-based curation. Plant Cell Physiol 56(1):e9–e9

Parry MAJ, Hawkesford MJ (2012) An integrated approach to crop genetic improvement. J Integr Plant Biol 54(4):250–259

Patti GJ, Yanes O, Siuzdak G (2012) Metabolomics: the apogee of the omics trilogy. Nat Rev Mol Cell Biol 13(4):263

Paupière MJ, van Heusden AW, Bovy AG (2014) The metabolic basis of pollen thermo-tolerance: perspectives for breeding. Meta 4(4):889–920

Poggio T, Girosi F (1990) Networks for approximation and learning. Proc IEEE 78(9):1481–1497

Pravdova V, Walczak B, Massart DL (2002) A comparison of two algorithms for warping of analytical signals. Anal Chim Acta 456(1):77–92

Rai A, Saito K (2016) Omics data input for metabolic modeling. Curr Opin Biotechnol 37:127–134

Ramalingam A, Kudapa H et al (2015) Proteomics and metabolomics: two emerging areas for legume improvement. Front Plant Sci 6:1116

Riedelsheimer CA, Czedik-Eysenberg et al (2012) Genomic and metabolic prediction of complex heterotic traits in hybrid maize. Nat Genet 44:217

Roberts LD, Souza AL, Gerszten RE, Clish CB (2012) Targeted metabolomics. Curr Protoc Mol Biol 30(2):1–24

Roessner U, Wagner C, Kopka J, Trethewey RN, Willmitzer L (2000) Simultaneous analysis of metabolites in potato tuber by gas chromatography-mass spectrometry. Plant J 23(1):131–142

Roessner-Tunali U, Hegemann B, Lytovchenko A, Carrari F, Bruedigam C, Granot D, Fernie AR (2003) Metabolic profiling of transgenic tomato plants overexpressing hexokinase reveals that the influence of hexose phosphorylation diminishes during fruit development. Plant Physiol 133(1):84–99

Saito K, Matsuda F (2010) Metabolomics for functional genomics, systems biology, and biotechnology. Annu Rev Plant Biol 61(1):463–489

Samuelsson LM, Larsson DG (2008) Contributions from metabolomics to fish research. Mol BioSyst 4(10):974–979

Sato Y, Takehisa H et al (2013) RiceXPro version 3.0: expanding the informatics resource for rice transcriptome. Nucleic Acids Res 41(Database issue):D1206–D1213

Sauerschnig C, Doppler M et al (2018) Methanol generates numerous artifacts during sample extraction and storage of extracts in metabolomics research. Meta 8(1):1

Sauvage C, Segura V et al (2014) Genome-wide association in tomato reveals 44 candidate loci for fruit metabolic traits. Plant Physiol 165(3):1120–1132

Sawada Y, Hirai MY (2013) Integrated lc-ms/ms system for plant metabolomics. Comput Struct Biotechnol J 4(5):e201301011

Sawada Y, Nakabayashi R, Yamada Y, Suzuki M, Sato M, Sakata A, Akiyama K, Sakurai T, Matsuda F, Aoki T, Hirai MY, Saito K (2012) RIKEN tandem mass spectral database (ReSpect) for phytochemicals: a plant-specific MS/MS-based data resource and database. Phytochemistry 82:38–45

Scott IM, Vermeer CP et al (2010) Enhancement of plant metabolite fingerprinting by machine learning. Plant Physiol 153(4):1506–1520

Sekiyama Y, Okazaki K, Kikuchi J, Ikeda S (2017) NMR-based metabolic profiling of field-grown leaves from sugar beet plants harbouring different levels of resistance to Cercospora leaf spot disease. Metabolites 7(1):4. pii:E4

Shen H, Grung B, Kvalheim OM, Eide I (2001) Automated curve resolution applied to data from multi-detection instruments. Anal Chim Acta 446(1–2):311–326

Shitan N (2016) Secondary metabolites in plants: transport and self-tolerance mechanisms. Biosci Biotechnol Biochem 80(7):1283–1293

Shyur LF, Yang NS (2008) Metabolomics for phytomedicine research and drug development. Curr Opin Chem Biol 12(1):66–71

Simó C, Ibáñez C et al (2014) Metabolomics of genetically modified crops. Int J Mol Sci 15(10):18941–18966

Smith CA, O'Maille G, Want EJ, Qin C, Trauger SA, Brandon TR, Custodio DE, Abagyan R, Siuzdak G (2005) METLIN: a metabolite mass spectral database. Ther Drug Monit 27(6):747–751

Smith CA, Want EJ, O'Maille G, Abagyan R, Siuzdak G (2006) XCMS: processing mass spectrometry data for metabolite profiling using nonlinear peak alignment, matching, and identification. Anal Chem 78(3):779–787

Sugimoto M, Kawakami M, Robert M, Soga T, Tomita M (2012) Bioinformatics tools for mass spectroscopy-based metabolomic data processing and analysis. Curr Bioinforma 7(1):96–108

Taketo O, Farit AM et al (2010) Metabolomics of medicinal plants: the importance of multivariate analysis of analytical chemistry data. Curr Comput Aided Drug Des 6(3):179–196

Taylor J, King RD, Altmann T, Fiehn O (2002) Application of metabolomics to plant genotype discrimination using statistics and machine learning. Bioinformatics 18(suppl_2):S241–S248. The Plant Journal 54(5), 949–962

Tian H, Lam S et al (2016) Metabolomics, a powerful tool for agricultural research. Int J Mol Sci 17(11):1871

Tikunov Y, Lommen A, De Vos CR, Verhoeven HA, Bino RJ, Hall RD, Bovy AG (2005) A novel approach for nontargeted data analysis for metabolomics. Large-scale profiling of tomato fruit volatiles. Plant Physiol 139(3):1125–1137

Tohge T, de Souza LP et al (2014) Genome-enabled plant metabolomics. J Chromatogr B 966:7–20

Toubiana D, Fernie AR et al (2013) Network analysis: tackling complex data to study plant metabolism. Trends Biotechnol 31(1):29–36

Udayakumar M, Chandar DP, Arun N, Mathangi J, Hemavathi K, Seenivasagam R (2012) PMDB: plant metabolome database—a metabolomic approach. Med Chem Res 21(1):47–52

van der Hooft JJJ, Vervoort J, Bino RJ, de Vos CH (2012) Spectral trees as a robust annotation tool in LC–MS based metabolomics. Metabolomics 8(4):691–703

Viant MR (2008) Recent developments in environmental metabolomics. Mol BioSyst 4(10):980–986

Vinayavekhin N, Saghatelian A (2010) Untargeted metabolomics. Curr Protoc Mol Biol Chapter 30:Unit 30.1.1-24

Wanichthanarak K, Fahrmann JF et al (2015) Genomic, proteomic, and metabolomic data integration strategies. Biomark Insights 10(Suppl 4):1–6

Ward JL, Harris C, Lewis J, Beale MH (2003) Assessment of 1H NMR spectroscopy and multivariate analysis as a technique for metabolite fingerprinting of *Arabidopsis thaliana*. Phytochemistry 62(6):949–957

Weckwerth W (2003) Metabolomics in systems biology. Annu Rev Plant Biol 54(1):669–689

Wen W, Jin M et al (2018) An integrated multi-layered analysis of the metabolic networks of different tissues uncovers key genetic components of primary metabolism in maize. Plant J 93(6):1116–1128

Wettschereck D, Aha DW, Mohri T (1997) A review and empirical evaluation of feature weighting methods for a class of lazy learning algorithms. Artif Intell Rev 11(1–5):273–314

Wiklund S, Johansson E, Sjöström L, Mellerowicz EJ, Edlund U, Shockcor JP, Trygg J (2008) Visualization of GC/TOF-MS-based metabolomics data for identification of biochemically interesting compounds using OPLS class models. Anal Chem 80(1):115–122

Wu S, Alseekh S et al (2016) Combined use of genome-wide association data and correlation networks unravels key regulators of primary metabolism in *Arabidopsis thaliana*. PLoS Genet 12(10):e1006363

Wu S, Tohge T et al (2018) Mapping the arabidopsis metabolic landscape by untargeted metabolomics at different environmental conditions. Mol Plant 11(1):118–134

Wurtzel ET, Kutchan TM (2016) Plant metabolism, the diverse chemistry set of the future. Science 353(6305):1232

Zhang A, Sun H et al (2012) Modern analytical techniques in metabolomics analysis. Analyst 137(2):293–300

Zhou J, Liu H, Liu Y, Liu J, Zhao X, Yin Y (2016) Development and evaluation of a parallel reaction monitoring strategy for large-scale targeted metabolomics quantification. Anal Chem 88(8):4478–4486

Zierer J, Menni C et al (2015) Integration of 'omics' data in aging research: from biomarkers to systems biology. Aging Cell 14(6):933–944

Zivy M, Wienkoop S et al (2015) The quest for tolerant varieties: the importance of integrating "omics" techniques to phenotyping. Front Plant Sci 6:448

Chapter 8
Bioinformatics and Medicinal Plant Research: Current Scenario

Insha Zahoor, Amrina Shafi, Khalid Majid Fazili, and Ehtishamul Haq

Contents

8.1   Introduction.. 141
8.2   Bioinformatics Resources in Medicinal Plant Research................................... 143
8.3   Conclusion... 150
References.. 151

8.1 Introduction

Bioinformatics has become essentially a multidisciplinary field owing to amalgamation of several different branches of science, thereby utilizing their applications in data analysis, interpretation and management. It has evolved as a fast-growing branch of biosciences with its impact on almost every aspect of research. Due to its growing requirement in analysing the vast amounts of data generated in day-to-day research, it has become a cornerstone for basic science research involving genomics, metabolomics, pharmacogenomics, transcriptomics, metagenomics, proteomics, etc. It has undoubtedly accelerated the process of knowledge dissemination and interpretation through its multifarious tools, online/web servers, databases, algorithms, computational analysis, etc. (Gu and Chen 2013). Bioinformatics has remarkably influenced several phases involved in the process of drug designing such as drug assessment, target identification, screening, refinement, development and resistance (Katara 2013). However, it is quite surprising that commercially

I. Zahoor (✉) · K. M. Fazili
Bioinformatics Centre, University of Kashmir, Srinagar, Jammu and Kashmir, India

Department of Biotechnology, School of Biological Sciences, University of Kashmir, Srinagar, Jammu and Kashmir, India
e-mail: drinsha@uok.edu.in; fazili@kashmiruniversity.ac.in

A. Shafi · E. Haq
Department of Biotechnology, School of Biological Sciences, University of Kashmir, Srinagar, Jammu and Kashmir, India
e-mail: haq@kashmiruniversity.ac.in

© Springer Nature Switzerland AG 2019
K. R. Hakeem et al. (eds.), *Essentials of Bioinformatics, Volume III*,
https://doi.org/10.1007/978-3-030-19318-8_8

available drugs (approximately one-third) are mostly derived from medicinal plants which make it mandatory to exploit the unearthed potential of this area of research where potential application of high-throughput bioinformatics approaches has largely been ignored (Miller 2011; Strohl 2000).

Medicinal plants continue to offer innumerable chemical compounds in the form of herbal formulations or herbal medicines which have been in use for human applications since ancient times as they tend to be renewable and comparatively safer for use (Newman and Cragg 2016; Saxena et al. 2013). They tend to be valuable plant-derived products in natural form with significant therapeutic potential making them quite preferable options for drugs in different systems of medicines (Karunamoorthi et al. 2013; Pye et al. 2017). They have been finding increasing use in treating basic health problems ranging from common cold to various skin ailments due to which their demand has increased tremendously in the present pharmaceutical set-up (Cowan 1999; Patwardhan et al. 2005). Different parts of the medicinal plants have been utilized for medicinal uses by human beings (Chanda 2014) as they tend to contain several valuable primary and secondary metabolites with therapeutic potential (Agyare et al. 2013; Incarbone and Dunoyer 2013). At the same time, there is a pressing demand for medicinal plants from the pharmaceutical companies because of their low-cost availability and commercial as well as healing benefits in contemporary medicine, but due to the adoption of traditional manual methods for initial screening of plants and subsequent processing, this area of research is lagging behind due to which market requirements remain largely unmet (Leonti 2013).

Based on the ethnobotanical data, there is tremendous potential for new drug development from these medicinal plants as there is much more scope to explore about them (Clarkson et al. 2004; Fabricant and Farnsworth 2001). There are certain challenges in exploiting medicinal plants for synthetic drug development due to inherent limitations pertaining to this area of research, which mainly includes costly and labour-intensive conventional manual bioscreening approaches (DiMasi et al. 2003; Fabricant and Farnsworth 2001). To harness the potential of medicinal plants for drug discovery, bioinformatics is appearing to play a substantial role in several aspects ranging from computational approaches to high-throughput screening involving computer-aided drug design (Jorgensen 2004). Different bioinformatics approaches have been utilized for studying diverse aspects of plants ranging from genomics to metabolomics which include genes, proteins, expression profiling of genes, metabolic processes such as metabolic profiling, biomolecular annotation and validation (Champagne and Boutry 2013; Gu and Chen 2013; Lagunin et al. 2014; Li et al. 2015; Saito 2013; Yang et al. 2014; Wolfender et al. 2013). A large number of research studies are being carried out around the world to employ several biotechnological approaches for deeper understanding of the molecular mechanisms behind metabolite synthesis so as to amplify the production of potent and valuable medicinal plant products (Ritala et al. 2014).

Bioinformatics-driven era will have extensive consequences on the overall advancement of medicinal plant research in the upcoming decades. So far, only a small facet of this area has been explored, and therefore, there is an urgent need for interdisciplinary research on medicinal plants (Chakraborty 2018). As a result, this chapter highlights our endeavour in providing comprehensive summary on the current scenario of the important applications of bioinformatics resources in the field of medicinal plants, which could provide significant help in tapping the hidden potential of this golden treasure for the future drug designing. Several dimensions of bioinformatics are discussed wherein different computational methods are being exploited to derive valuable, effective and timely information from medicinal plants. We present a concise and updated summary on the utility and potential applications of commonly used bioinformatics tools and databases in driving the field of medicinal plant research by performing thorough analysis of data as well as integrating the scattered data.

8.2 Bioinformatics Resources in Medicinal Plant Research

There exists ginormous ethnobotanical data that could have high potential for serving as source of contemporary medicine. However, at the same time, this data is highly scattered and unlinked which certainly impedes the usefulness of the knowledge on medicinal plants. So, there has to be some automatic system for medicinal plant data storage, its organization, interpretation and extraction in a user-friendly format. Bioinformatics in this context has a crucial role to play as it provides tools to handle segregated data generated from high-throughput "omics" technologies that certainly aid in extraction of meaningful information and lead to rapid drug discovery and development from medicinal plants (Babar et al. 2017). Figure 8.1 provides the schematics of the role of bioinformatics in integrating "omics approaches"

Fig. 8.1 The application of bioinformatics on the data integration and analysis generated from "omics" approaches of medicinal plants aids in quicker discovery and generation of novel herbal drugs

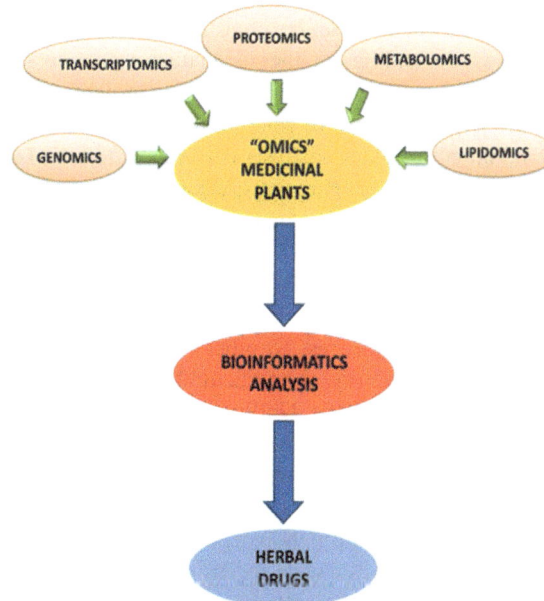

in the area of medicinal plant research. Data mining approaches can be used for introducing bioprospecting and screening of medicinal plants (Harishchander 2017; Sharma and Sarkar 2012). So far, there are very less number of plants whose whole genome sequence data is available, and the first attempt to digitize plant-based data comes from the Biodiversity Heritage Library (http://www.biodiversitylibrary.org/) and the China-US Million Book Digital Library Project (http://www.cadal.zju.edu.cn). However, there were certain challenges inherent to these libraries in terms of data extraction from them. Later on, with the usage of automated algorithms, data extraction could be simplified by applying these advanced approaches on the historical texts, and one such application was based on herbal text of the seventeenth century, *Ambonese Herbal* (Buenz et al. 2005). However, at the same time, there are certain issues in analysing the historic texts especially due to diversity in taxonomic recognition of the plants across different historical texts owing to the usage of different names and the languages or scripts in the text (Sarkar 2000; Wagner et al. 1998). To circumvent such issues, Samwald et al. proposed a semantic-based approach for clubbing scattered and diverse data sources (Samwald et al. 2010). Based on similar lines, several valuable compounds from Chinese medicine have been found to have pharmaceutical value in treating a wide continuum of psychological problems (Samwald et al. 2010). Another similar approach based on such technology for large-scale data integration is the semantic e-Science infrastructure for Traditional Chinese Medicine (TCM) system (Chen et al. 2007).

The primary bioinformatics approach for the study of medicinal plants is through development of easily accessible databases and programs by different government and private research organizations through international collaborations as a part of consortia. These databases are being maintained and updated on a regular basis. Some databases provide exclusive information on medicinal plants like commonly used literature databases etc. (Sayers et al. 2010). Another category of databases is community-based depending on the molecular aspect to be explored such as studying molecular networks, pathway mapping, protein interactions, metabolic pathways, genome sequence analysis, gene expression, subcellular localization, domain identification and comparative analysis between different plant species (Babar et al. 2017; Lagunin et al. 2014). There are specific databases also available on medicinal plants, although information can even be retrieved through primary bioinformatics resources including National Centre for Biotechnology Information (NCBI) at the US National Library of Medicine (NLM) (Sayers et al. 2010), Kyoto Encyclopedia of Genes and Genomes (KEGG) (Kanehisa and Goto 2000) and KNApSAcK (Afendi et al. 2012). These provide easy access to other databases like GenBank (Benson et al. 2013), Research Collaboratory for Structural Bioinformatics (RCSB)-Protein Data Bank (PDB) (Rose et al. 2013), UniProt (Magrane and Consortium 2011; UniProt 2014) and Gene Expression Omnibus (GEO) (Edgar et al. 2002). Further, there are several other databases which provide exclusive information on medicinal plants. One such non-commercial database on ethnobotanical data is International Ethnobotany Database (ebDB) (http://ebdb.org). It provides data retrieval options in different languages and exhibits a wide spectrum of options for accessing complete information from location to robust searching with unique data export options.

Furthermore, there are several databases serving the purpose of data analysis for medicinal plants and their products, functions, composition, adverse effects including toxicity as well as their therapeutic applications. One such example is NAPALERT (Loub et al. 1985), Dr. Park's United States Department of Agriculture (USDA) database (http://www.pl.barc.usda.gov/usda_info/disease_intro.cfm?id=39) and Dr. Duke's Phytochemical and Ethnobotanical Database at USDA (http://www.ars-grin.gov/duke/). Another comprehensive compilation on medicinal plants is TCM Information Database (TCM-ID) (Ji et al. 2006; Wang et al. 2005; Xue et al., 2013) and Herb Ingredient's Target (HIT) (Ye et al. 2011). HIT has been developed for providing therapeutic targets to the potential herbal leads from TCM. It provides links to several other resources like NCBI (Sayers et al. 2010), PDB (Bernstein et al. 1978), UniProt (Magrane and Consortium 2011; UniProt 2014), Pfam (Punta et al. 2012), Therapeutic Target Database (TTB) (Liu et al. 2011), TCM-ID (Ji et al. 2006; Wang et al. 2005; Xue et al. 2013) and KEGG (Kanehisa and Goto 2000). For more specific information, text mining is being employed in creation of databases like MedMiner, PathBinder and PreBIND (Donaldson et al. 2003; Tanabe et al. 1999; Zhang et al. 2009) and developing gene ontology (GO) annotation platforms (Consortium 2013). There exist several other databases based on the geographic distribution of medicinal plants, thus making them region specific like Australian medicinal knowledge base CMKb (Gaikwad et al. 2008). It provides users with different information modules linked to several sources including NCBI and PubChem. Another recent approach in this direction has been the creation of manually curated comprehensive openly accessible online database exclusively for Indian medicinal plants called Indian Medicinal Plants Phytochemicals and Therapeutics (IMPPAT) (Mohanraj et al. 2018). It has been established with the help of cheminformatic approaches to provide a platform on the phytochemistry of 1742 medicinal plants from Indian origin, particularly from Himalayan region. It will serve as harbinger for plant-based drug discovery through in silico approaches. Another similar resource is Raintree (http://rain-tree.com/ethnic.htm) database providing exclusive information on medicinal plants habituating Amazon rainforest.

After literature study on different aspects of medicinal plants, the molecular study at different "omics" levels follows, which includes genomics, transcriptomics, proteomics and metabolomics. These approaches can become more advanced and efficient due to the latest myriad of interventions in bioinformatics field for data analysis. The first intervention of bioinformatics at the molecular level is through full-length genome sequencing involving next-generation sequencing approaches (NGS), and there are several databases which store plant-specific full-length genomes (Benson et al. 2013). Most of the genomic information is managed and analysed through expressed sequence tags (EST) (Ueno et al. 2012). After sequence analysis, a lot of bioinformatics approaches are being used for the sequence assembly, and later on comparative analysis of the sequences is being performed with the help of bioinformatics softwares (Drezen and Lavenier 2014). These techniques provide cost- and time-effective huge volumes of genomic data. Other latest sequencing techniques based on advanced bioinformatics suites include comparative hybridization of probes, polymorphism ratio sequencing and the 454 method, all of which require very less quantity (in picolitre) of sample to be sequenced (Li and Quiros 2001; Schena et al. 1995; van Dijk et al. 2014). Sequence analysis provides clues for further validation of data at mRNA level, protein and phylogenetics (Caetano-Anollés and Gresshoff 2013). However, sequence analysis is prone to errors since it depends on the algorithms used in developing a particular bioinformatics program. There are several other programs based on advanced computational algorithms available for studying genetic elements associated with medicinal plants including exonic and intronic sequences which could be actually exploited for identification of genetic elements involved in secondary metabolite production as well as gene characterization of medicinal plants for their propagation (Babar et al. 2017). After sequence assembly, sequence annotation is done using several platforms like KEGG (Kanehisa and Goto 2000), SwissProt (Boeckmann et al. 2003), The Arabidopsis Information Resource (TAIR) (Berardini et al. 2015) and nucleotide databases at NCBI (Sharma and Sarkar 2012). Annotation of sequenced data is exclusively done for the detection of genome-specific signatures, which includes some repetitive sequences such as simple sequence repeat (SSR) markers (Childs 2014; Tatusova et al. 2016). The identification and characterization of SSRs are carried out using computational platform of SSRLocator (da Maia et al. 2008). Molecular marker recognition helps in establishing linkage studies, evolutionary relationship, comparative genomics, gene function prediction and genome organization (Davies et al. 2013; Varshney et al. 2005).

Most of the genomic and transcriptomic resources are available at NCBI through Plant Genome Central; however, EST-related resources provide a major drawback as the available data is of poor quality and present in highly unorganized form (Sharma and Sarkar 2012). On the other hand, efficient transcriptomic analysis of few plant species is performed on the platform of EGENES database by clubbing together EST-based genomic data with information at RNA level (Masoudi-Nejad et al. 2007). Another similar platform providing exclusive transcriptome and metabolome information for medicinal plants is the Medicinal Plant Genomics Resource (MPGR) (http://medicinalplantgenomics.msu.edu/) maintained by the Buell Lab in the Plant Biology Department at Michigan State University, but it provides information pertaining to taxonomically different species only. Transcriptome analysis is useful in determining the regulatory sequences and networks, in identifying transcription factors and their role, in determining the activity of intronic sequences and also in deciphering the function and expression of medicinal plant proteins at transcript level (Dhondt et al. 2013). The primary molecular biology techniques involved in studying transcriptome of medicinal plants include DNA microarray (Singh and Kumar 2013), whole genome array (WGA) (Xu et al. 2012) and immunoprecipitation (Ren et al. 2000), and data generated thereafter is thoroughly analysed using different bioinformatics programs. There are several medicinal plant-based resources available for analysing the gene expression data such as Plant Expression Database (PLEXdb) (Dash et al. 2012; Wise et al. 2007), GEO (Barrett and Edgar 2006) and EBI ArrayExpress (Rocca-Serra et al. 2003). Bioinformatics suites provide a vast range of options for data retrieval and employ a complex exhaustive computational and statistical approach for establishing the networks and pathways (Babar et al. 2017). Bioinformatics approaches have even been used to generate co-expression networks as has been done with barley and triticale for revealing regulatory networks related to drought stress and cellulose biosynthesis (Mochida et al. 2011). Other tools involved in comparing transcriptomics data generated from different experiments are Affycomp (Irizarry et al. 2006) and Bioconductor (Durinck et al. 2005). Furthermore, there are several meta-analysis databases and programs available for profile comparison of medicinal plant genes (Hegde et al. 2000; Engelhorn and Turck 2017). Table 8.1 provides an overview of bioinformatics tools and databases available for the study of medicinal plants.

Moreover, the outcomes from proteomics studies of medicinal plants ranging from determination of the protein structure and function to interacting partners and subcellular localization are being analysed and interpreted with the help of bioinformatics approaches (Wetie et al. 2014). Presently, more advanced and refined computational tools are being developed for understanding proteomics data to enable better analysis of plant proteins. There are some tools meant for thorough evaluation of gel images at the level of gel electrophoresis (Caccia et al. 2013). Similarly, there are yet other tools meant for analysing and interpreting high-throughput data generated through mass spectrometry (Bensimon et al. 2012). These tools are used for predicting better outcomes as well as molecular weight of the characterized proteins based on probability score. However, the overall output has low confidence value due to inherent program errors, which make computational evaluations relatively error prone. Variations at the level of proteomics data can have significant impact on the function of the protein, which necessitates the need for collaborating proteomics and bioinformatics data for better understanding of the functionality (Yang et al. 2013). One such freely accessible manually curated database is TarNet, which provides in-depth network construction analysis of biological pathways and aids in deciphering protein–protein interactions and plant–compound–protein relationships (Hu et al. 2016). The proteomics data can also be used for molecular docking approaches involved in the process of drug designing and discovery. The docking approaches are employed for studying kinetic and stability parameters of protein–ligand complex formation between phytochemicals and other proteins (London et al. 2013). Another crucial application of bioinformatics can be seen at the level of plant metabolomics studies wherein computational approaches are used for data interpretation which can be used later on for the generation of correlation networks (Kempinski et al. 2015). One such approach has been the genome-wide prediction of metabolic enzymes, pathways and gene clusters in plants through Plant Metabolic Network (PMN) (Schläpfer et al. 2017). Metabolic profiling is very crucial step for a prospective phytochemical before entering into pharmacological screening stage. Bioinformatics approaches for studying metabolomics are quite complex as they are based on the stoichiometry parameters. However, they are being modelled to address the complexity issues of large metabolic networks so as to predict the localization pattern of a particular metabolite by mimicking the in vivo cellular conditions (Boudon et al. 2015).

Interestingly, there is another approach for determining gene interaction and their possible roles through pathway analysis of plant secondary metabolites using

Table 8.1 Overview of bioinformatics resources available for studying medicinal plants

Literature	Genomics	Transcriptomics	Proteomics	Metabolomics
NAPALERT Loub et al. (1985)	KEGG Kanehisa and Goto (2000)	EBI ArrayExpress Rocca-Serra et al. (2003)	PDB Bernstein et al. (1978)	METLIN Smith et al. (2005)
HerbMed Wootton (2002)	VISTA Mayor et al. (2000)	miRBase Griffiths-Jones et al. (2006)	SwissProt Boeckmann et al. (2003)	MetaCyc Caspi et al. (2008)
PubMed Wheeler et al. (2007)	MPGR http://medicinalplantgenomics.msu.edu/	EGENES Masoudi-Nejad et al. (2007)	UniProt Magrane and Consortium (2011) and UniProt (2014)	PathPred Moriya et al. (2010)
TCM-ID Wang et al. (2005), Ji et al. (2006), and Xue et al. (2013)	Ensembl Plants http://plants.ensembl.org/index/	PLEXdb Dash et al. (2012)	RCBS Rose et al. (2013)	KNApSAcK Afendi et al. (2012)
HIT Ye et al. (2011)	Plant GDB Genome Browser http://www.plantgdb.org/	Plant Omics Data Center Ohyanagi et al. (2014)	TarNet Hu et al. (2016)	PMN Schläpfer et al. (2017)

platform of KEGG (Kanehisa and Goto 2000; Kanehisa et al. 2010). KEGG Drug database provides the crucial information on molecular networks including biosynthetic pathways involved in the production of bioactive molecules within the plants, and it also shows the interaction between prospective drug compounds and their possible targets (Kanehisa 2009). It contains the structures of drug compounds from over-the-counter (OTC) drugs and TCM (Kanehisa et al. 2010; Kanehisa 2009). Another similar platform like KEGG is PathPred, which is a web server exclusively used for predicting common pathways associated with structurally related compounds (Moriya et al. 2010). There are some other computational approaches for drug discovery from unexplored medicinal plants. Some of the examples include TCM@Taiwan (Chen 2011), Cardiovascular Disease Herbal Database (CVDHD) (Gu et al. 2013), Naturally occurring Plant-based Anti-cancer Compound-activity-Target database (NPACT) (Mangal et al. 2013), NutriChem (Jensen et al. 2015), Phytochemica (Pathania et al. 2015), TCM-Mesh (Zhang et al. 2017) and Natural Product Activity and Species Source database (NPASS) (Zeng et al. 2018). The ultimate role of bioinformatics approaches in studying medicinal plants is reflected through ethnopharmacology at the level of drug discovery. The computational approaches have dramatically resulted in quicker drug discovery owing to virtual screening. It is very difficult to establish a particular property and or activity for whole plant extracts as they are known to contain more than one constituent. There has to be some optimization and rationalization of herbal drugs, which can be addressed through application of computational techniques. There are two approaches in this direction, Quantitative Composition-Activity Relationship (QCAR) (Zhao et al. 2004) and Quantitative Structure–Activity Relationship (QSAR) (Nantasenamat et al. 2009), however the former one being commonly used in drug designing. In this manner, biological and pharmacological activity of herbal drugs can be predicted quantitatively based on the composition and structure of the potent drug compounds. These approaches can be effectively optimized and modelled to improve overall accuracy of their computational predictions by applying machine-learning techniques including Genetic Algorithm-Artificial Neural Network, Multiple Linear Regression, Artificial Neural Network and Support Vector Machines (Nayak et al. 2010; Wang et al. 2006).

Before directly testing drugs (plant products-natural, synthetic and semi-synthetic) on humans during clinical trials, preclinical testing in cell lines and animal models is carried out for establishing their biochemical properties and therapeutic potential of compounds as well as checking their toxicity and adverse effects which is really an expensive and time-consuming process (Babar et al. 2017). Here, bioinformatics resources are employed for correlating "omics" data with drug intervention response. There has been a tremendous progress in the development of medicinal plant databases with the prime focus on the phytochemistry of natural plant products, key active plant metabolites and their constituents, pharmacokinetics of potential herbal drug compounds, drug interactions, structural information, remedial applications, toxicity and adverse effects, all of which can certainly provide a platform for high-throughput virtual screening of prospective drug compounds (Chakraborty 2018; Jensen et al. 2014; Mangal et al. 2013). The application of computational methods can yield in better understanding of the pharmacokinetic and pharmacodynamic aspects of phytochemical drugs. In the process of drug screening and compound characterization through systems biology, multipathway approach should be employed against high-throughput virtual screening, which considers only one pathway as former approach and enables biological activity-based classification of the hits and leads (Pei et al. 2014).

Furthermore, toxicological and safety evaluations of drug compounds form a crucial part of herbal drug development and standardization process (Wu et al. 2004). In fact, there have been very limited studies on toxicological profile of plant-based drugs as it is a quite time-consuming and expensive process (Wu et al. 2004). However, computational toxicology assessment approaches are offering better means by providing cost-effective modelling and prediction options for eliminating toxic natural products at early stages of screening (Rusyn and Daston 2010; Valerio et al. 2010). In this direction, two QSAR-based tools (LMA and MC4PC) have been used for testing carcinogenicity of plant products in rodents by comparing with standard phytochemicals whose carcinogenicity was known (Valerio et al. 2010; Yang et al. 2008). Apart from this, toxicogenomics and statistical learning approaches have also been employed in toxicology assessment, as they are more reliable because they depend only on basic structural features and physiochemical properties of compounds (Li et al. 2005; Youns et al. 2010). In this way, bioinformatics can really transform medicinal plant research by primarily governing drug development process through computationally derived hypothesis involving exhaustive algorithms and statistical methods, thus making drug designing and discovery much better in terms of time and money.

8.3   Conclusion

Medicinal plants offer a great deal of benefits in the form of herbal drugs to mankind; however, research in this field encounters many challenges due to expensive, slow and tedious conventional approaches being adopted in plant-based drug discovery. Even if there are huge volumes of data available on medicinal plants, data remains highly dispersed and unlinked which certainly hinders the drug development process as it depreciates the information pertaining to potential drug sources and bioactive metabolites. Nonetheless, to increase the potential of research in the area of medicinal plants, bioinformatics has come to rescue as it has enabled application of high-throughput computational tools and approaches for plant-based drug discovery which has particularly lead to discovery of new genes, different omics pathways and networks pertaining to bioactive metabolite production in medicinal plants. Bioinformatics in the present era has become the backbone of biological sciences and has revolutionized different aspects of biological research. It has evolved as a fast-growing branch of biosciences with its impact on almost every aspect of research.

In the present chapter, we attempted to put forth a comprehensive but updated summary on the role of different bioinformatics approaches in studying medicinal plants. This has really revolutionized the area of medicinal plant research by automation of plant-based drug discovery process. Bioinformatics to some extent has acted as a bridge by integrating scattered data and converting it to useful information. It has more or less developed more refined and targeted approaches for medicinal plant-based searches. The interaction of "omics" technologies with bioinformatics would definitely assist in exploring plant systems in a better way as it can provide us controlled information about several pathways involved in the synthesis of different valuable plant metabolites. Later on, these pathways can also be manoeuvred as per human requirements in terms of production of pharmacologically valuable compounds. There are certain limitations inherent to the field of bioinformatics, yet it has provided a platform for significant progress in the area of medicinal plant research. So far, there has been no approach for exploring genotype-phenotype correlations. However, it still continues to provide a whole lot of benefits in the field of medicinal plant research. To get an in-depth understanding of the plant-based mechanisms behind various cellular processes, bioinformatics can be helpful by providing information on plant genomics, transcriptomics, proteomics and metabolomics through an array of computational tools, thereby facilitating identification of novel plant sources for future drug development.

Acknowledgements We are highly indebted to the Bioinformatics Centre, University of Kashmir, for providing their services while drafting this chapter.

Conflict of Interest The authors declare that they have no conflicts of interest with respect to research, authorship and publication of this book chapter.

References

Afendi FM, Okada T, Yamazaki M, Hirai-Morita A, Nakamura Y, Nakamura K, Ikeda S, Takahashi H, Altaf-Ul-Amin M, Darusman LK, Saito K, Kanaya S (2012) KNApSAcK family databases: integrated metabolite-plant species databases for multifaceted plant research. Plant Cell Physiol 53(2):e1

Agyare C, Obiri DD, Boakye YD, Osafo N (2013) Anti-inflammatory and analgesic activities of African medicinal plants. In: Kuete V (ed) Medicinal plant research in Africa. Elsevier, London, pp 725–752

Babar MM, Najam-us-Sahar ZS, Pothineni RV, Ali Z, Faisal S, Hakeem RK, Gul A (2017) Application of bioinformatics and system biology in medicinal plant studies. In: Hakeem KR et al (eds) Plant bioinformatics. Springer International Publishing, Cham, pp 375–393

Barrett T, Edgar R (2006) Gene expression omnibus: microarray data storage, submission, retrieval, and analysis. Methods Enzymol 411:352–369

Bensimon A, Heck AJ, Aebersold R (2012) Mass spectrometry-based proteomics and network biology. Annu Rev Biochem 81:379–405

Benson DA, Cavanaugh M, Clark K, Karsch-Mizrachi I, Lipman DJ, Ostell J, Sayers EW (2013) GenBank. Nucleic Acids Res 41(D1):D36–D42

Berardini TZ, Reiser L, Li D, Mezheritsky Y, Muller R, Strait E, Huala E (2015) The Arabidopsis information resource: making and mining the "gold standard" annotated reference plant genome. Genesis 53(8):474–485

Bernstein FC, Koetzle TF, Williams GJ, Meyer EF Jr, Brice MD, Rodgers JR, Kennard O, Shimanouchi T, Tasumi M (1978) The protein data bank: a computer-based archival file for macromolecular structures. Arch Biochem Biophys 185(2):584–591

Boeckmann B, Bairoch A, Apweiler R, Blatter M-C, Estreicher A, Gasteiger E, Martin MJ, Michoud K, O'Donovan C, Phan I (2003) The SWISS-PROT protein knowledgebase and its supplement TrEMBL in 2003. Nucleic Acids Res 31(1):365–370

Boudon F, Chopard J, Ali O, Gilles B, Hamant O, Boudaoud A, Traas J, Godin C (2015) A computational framework for 3D mechanical modeling of plant morphogenesis with cellular resolution. PLoS Comput Biol 11(1):e1003950

Buenz EJ, Johnson HE, Beekman EM, Motley TJ, Bauer BA (2005) Bioprospecting Rumphius's Ambonese herbal: volume I. J Ethnopharmacol 96(1–2):57–70

Caccia D, Dugo M, Callari M, Bongarzone I (2013) Bioinformatics tools for secretome analysis. Biochim Biophys Acta (BBA) Protein Proteomics 1834(11):2442–2453

Caetano-Anollés G, Gresshoff P (2013) Phylogenetic analysis of plants. Mol Ecol Evol Approaches Appl 69:17

Caspi R, Foerster H, Fulcher CA, Kaipa P, Krummenacker M, Latendresse M, Paley S, Rhee SY, Shearer AG, Tissier C (2008) The MetaCyc database of metabolic pathways and enzymes and the BioCyc collection of pathway/genome databases. Nucleic Acids Res 36(suppl 1):D623–D631

Chakraborty P (2018) Herbal genomics as tools for dissecting new metabolic pathways of unexplored medicinal plants and drug discovery. Biochim Open 6:9–16

Champagne A, Boutry M (2013) Proteomics of nonmodel plant species. Proteomics 13(3–4):663–673

Chanda S (2014) Importance of pharmacognostic study of medicinal plants: an overview. J Pharmacogn Phytochem 2(5):69–73

Chen CY (2011) TCM Database@Taiwan: the world's largest traditional Chinese medicine database for drug screening in silico. PLoS One 6(1):e15939

Chen H, Mao Y, Zheng X, Cui M, Feng Y, Deng S, Yin A, Zhou C, Tang J, Jiang X, Wu Z (2007) Towards semantic e-science for traditional Chinese medicine. BMC Bioinforma 8(Suppl 3):S6

Childs KL (2014) Methods for plant genome annotation. In: Bell E (ed) Molecular life sciences: an encyclopedia reference. Springer, New York, pp 1–7

Clarkson C, Maharaj VJ, Crouch NR, Grace OM, Pillay P, Matsabisa MG, Bhagwandin N, Smith PJ, Folb PI (2004) In vitro antiplasmodial activity of medicinal plants native to or naturalized in South Africa. J Ethnopharmacol 92(2–3):177–191

Consortium GO (2013) Gene ontology annotations and resources. Nucleic Acids Res 41(D1):D530–D535

Cowan MM (1999) Plant products as antimicrobial agents. Clin Microbiol Rev 12(4):564–582

da Maia LC, Palmieri DA, de Souza VQ, Kopp MM, de Carvalho FI, Costa de Oliveira A (2008) SSR locator: tool for simple sequence repeat discovery integrated with primer design and PCR simulation. Int J Plant Genomics 2008:412696

Dash S, Van Hemert J, Hong L, Wise RP, Dickerson JA (2012) PLEXdb: gene expression resources for plants and plant pathogens. Nucleic Acids Res 40(D1):D1194–D1201

Davies TJ, Wolkovich EM, Kraft NJ, Salamin N, Allen JM, Ault TR, Betancourt JL, Bolmgren K, Cleland EE, Cook BI (2013) Phylogenetic conservatism in plant phenology. J Ecol 101(6):1520–1530

Dhondt S, Wuyts N, Inzé D (2013) Cell to whole-plant phenotyping: the best is yet to come. Trends Plant Sci 18(8):428–439

DiMasi JA, Hansen RW, Grabowski HG (2003) The price of innovation: new estimates of drug development costs. J Health Econ 22(2):151–185

Donaldson I, Martin J, De Bruijn B, Wolting C, Lay V, Tuekam B, Zhang S, Baskin B, Bader GD, Michalickova K (2003) PreBIND and textomy–mining the biomedical literature for protein-protein interactions using a support vector machine. BMC Bioinforma 4(1):1

Drezen E, Lavenier D (2014) Quality metrics for benchmarking sequences comparison tools. In: Brazilian symposium on bioinformatics. Springer International Publishing, Cham, pp 144–153

Durinck S, Moreau Y, Kasprzyk A, Davis S, De Moor B, Brazma A, Huber W (2005) BioMart and bioconductor: a powerful link between biological databases and microarray data analysis. Bioinformatics 21(16):3439–3440

Edgar R, Domrachev M, Lash AE (2002) Gene expression omnibus: NCBI gene expression and hybridization array data repository. Nucleic Acids Res 30(1):207–210

Engelhorn J, Turck F (2017) Meta-analysis of genome-wide chromatin data. Methods Mol Biol 1456:33

Fabricant DS, Farnsworth NR (2001) The value of plants used in traditional medicine for drug discovery. Environ Health Perspect 109(Suppl 1):69–75

Gaikwad J, Khanna V, Vemulpad S, Jamie J, Kohen J, Ranganathan S (2008) CMKb: a web-based prototype for integrating Australian aboriginal customary medicinal plant knowledge. BMC Bioinforma 9(Suppl 12):S25

Griffiths-Jones S, Grocock RJ, Van Dongen S, Bateman A, Enright AJ (2006) miRBase: microRNA sequences, targets and gene nomenclature. Nucleic Acids Res 34(Suppl 1):D140–D144

Gu P, Chen H (2013) Modern bioinformatics meets traditional Chinese medicine. Brief Bioinform 15(6):984–1003

Gu J, Gui Y, Chen L, Yuan G, Xu X (2013) CVDHD: a cardiovascular disease herbal database for drug discovery and network pharmacology. J Cheminform 5(1):51

Harishchander A (2017) A review on application of bioinformatics in medicinal plant research. Bioinforma Proteomics Open Access J 1(1):000104

Hegde P, Qi R, Abernathy K, Gay C, Dharap S, Gaspard R, Hughes J, Snesrud E, Lee N, Quackenbush J (2000) A concise guide to cDNA microarray analysis. BioTechniques 29(3):548–563

Hu R, Ren G, Sun G, Sun X (2016) TarNet: an evidence-based database for natural medicine research. PLoS One 11(6):e0157222

Incarbone M, Dunoyer P (2013) RNA silencing and its suppression: novel insights from in planta analyses. Trends Plant Sci 18(7):382–392

Irizarry RA, Cope LM, Wu Z (2006) Feature-level exploration of a published Affymetrix GeneChip control dataset. Genome Biol 7(8):1

Jensen K, Panagiotou G, Kouskoumvekaki I (2014) Correction: integrated text mining and che-moinformatics analysis associates diet to health benefit at molecular level. PLoS Comput Biol 10(1):10

Jensen K, Panagiotou G, Kouskoumvekaki I (2015) NutriChem: a systems chemical biology resource to explore the medicinal value of plant-based foods. Nucleic Acids Res 43(D1):D940–D945

Ji ZL, Zhou H, Wang JF, Han LY, Zheng CJ, Chen YZ (2006) Traditional Chinese medicine information database. J Ethnopharmacol 103(3):501

Jorgensen WL (2004) The many roles of computation in drug discovery. Science 303(5665):1813–1818

Kanehisa M (2009) Representation and analysis of molecular networks involving diseases and drugs. Genome Inform 23(1):212–213

Kanehisa M, Goto S (2000) KEGG: kyoto encyclopedia of genes and genomes. Nucleic Acids Res 28(1):27–30

Kanehisa M, Goto S, Furumichi M, Tanabe M, Hirakawa M (2010) KEGG for representation and analysis of molecular networks involving diseases and drugs. Nucleic Acids Res 38(D1):D355–D360

Karunamoorthi K, Jegajeevanram K, Vijayalakshmi J, Mengistie E (2013) Traditional medicinal plants a source of phytotherapeutic modality in resource-constrained health care settings. J Evid Based Complement Alternat Med 18(1):67–74

Katara P (2013) Role of bioinformatics and pharmacogenomics in drug discovery and development process. Netw Model Anal Health Inform Bioinforma 2(4):225–230

Kempinski C, Jiang Z, Bell S, Chappell J (2015) Metabolic engineering of higher plants and algae for isoprenoid production. In: Biotechnology of isoprenoids. Springer International Publishing, Cham, pp 161–199

Lagunin AA, Goel RK, Gawande DY, Pahwa P, Gloriozova TA, Dmitriev AV, Ivanov SM, Rudik AV, Konova VI, Pogodin PV, Druzhilovsky DS, Poroikov VV (2014) Chemo- and bioinformatics resources for in silico drug discovery from medicinal plants beyond their traditional use: a critical review. Nat Prod Rep 31(11):1585–1611

Leonti M (2013) Traditional medicines and globalization: current and future perspectives in ethnopharmacology. Front Pharmacol 4:92

Li G, Quiros CF (2001) Sequence-related amplified polymorphism (SRAP), a new marker system based on a simple PCR reaction: its application to mapping and gene tagging in Brassica. Theor Appl Genet 103(2–3):455–461

Li H, Ung CY, Yap CW, Xue Y, Li ZR, Cao ZW, Chen YZ (2005) Prediction of genotoxicity of chemical compounds by statistical learning methods. Chem Res Toxicol 18:1071–1080

Li X, Yang Y, Henry RJ, Rossetto M, Wang Y, Chen S (2015) Plant DNA barcoding: from gene to genome. Biol Rev 90(1):157–166

Liu X, Zhu F, Ma X, Tao L, Zhang J, Yang S, Wei Y, Chen YZ (2011) The therapeutic target database: an internet resource for the primary targets of approved, clinical trial and experimental drugs. Expert Opin Ther Targets 15(8):903–912

London N, Raveh B, Schueler-Furman O (2013) Druggable protein–protein interactions–from hot spots to hot segments. Curr Opin Chem Biol 17(6):952–959

Loub WD, Farnsworth NR, Soejarto DD, Quinn ML (1985) NAPRALERT: computer handling of natural product research data. J Chem Inf Comput Sci 25(2):99–103

Magrane M, Consortium U (2011) UniProt knowledgebase: a hub of integrated protein data. Database J Biol Databases Curation 2011:bar009

Mangal M, Sagar P, Singh H, Raghava GP, Agarwal SM (2013) NPACT: naturally occurring plant-based anti-cancer compound-activity-target database. Nucleic Acids Res 41(D1):D1124–D1129

Masoudi-Nejad A, Goto S, Jauregui R, Ito M, Kawashima S, Moriya Y, Endo TR, Kanehisa M (2007) EGENES: transcriptome-based plant database of genes with metabolic pathway information and expressed sequence tag indices in KEGG. Plant Physiol 144(2):857–866

Mayor C, Brudno M, Schwartz JR, Poliakov A, Rubin EM, Frazer KA, Pachter LS, Dubchak I (2000) VISTA: Visualizing global DNA sequence alignments of arbitrary length. Bioinformatics 16(11):1046–1047

Miller J (2011) The discovery of medicines from plants: a current biological perspective. Econ Bot 65:396–407

Mochida K, Uehara-Yamaguchi Y, Yoshida T, Sakurai T, Shinozaki K (2011) Global landscape of a co-expressed gene network in barley and its application to gene discovery in Triticeae crops. Plant Cell Physiol 52(5):785–803

Mohanraj K, Karthikeyan BS, Vivek-Ananth RP, Chand RPB, Aparna SR, Mangalapandi P, Samal A (2018) IMPPAT: a curated database of Indian medicinal plants, phytochemistry and therapeutics. Sci Rep 8(1):4329

Moriya Y, Shigemizu D, Hattori M, Tokimatsu T, Kotera M, Goto S, Kanehisa M (2010) PathPred: an enzyme-catalyzed metabolic pathway prediction server. Nucleic Acids Res 38(W1):W138–W143

Nantasenamat C, Isarankura-Na-Ayudhya C, Naenna T, Prachayasittikul V (2009) A practical overview of quantitative structure-activity relationship. EXCLI J 8:74–88

Nayak SK, Patra PK, Padhi P, Panda A (2010) Optimization of herbal drugs using soft computing approach. Int J Log Comput 1:34–39

Newman DJ, Cragg GM (2016) Natural products as sources of new drugs from 1981 to 2014. J Nat Prod 79(3):629–661

Ohyanagi H, Takano T, Terashima S, Kobayashi M, Kanno M, Morimoto K, Kanegae H, Sasaki Y, Saito M, Asano S (2014) Plant omics data center: an integrated web repository for interspecies gene expression networks with NLP-based curation. Plant Cell Physiol 56(1):e9. (1–8)

Pathania S, Ramakrishnan SM, Bagler G (2015) Phytochemica: a platform to explore phytochemicals of medicinal plants. Database (Oxford) 2015

Patwardhan B, Warude D, Pushpangadan P, Bhatt N (2005) Ayurveda and traditional Chinese medicine: a comparative overview. Evid Based Complement Alternat Med 2(4):465–473

Pei J, Yin N, Ma X, Lai L (2014) Systems biology brings new dimensions for structure-based drug design. J Am Chem Soc 136(33):11556–11565

Punta M, Coggill PC, Eberhardt RY, Mistry J, Tate J, Boursnell C, Pang N, Forslund K, Ceric G, Clements J, Heger A, Holm L, Sonnhammer EL, Eddy SR, Bateman A, Finn RD (2012) The Pfam protein families database. Nucleic Acids Res 40:D290–D301

Pye CR, Bertin MJ, Lokey RS, Gerwick WH, Linington RG (2017) Retrospective analysis of natural products provides insights for future discovery trends. Proc Natl Acad Sci USA 114(22):5601–5606

Ren B, Robert F, Wyrick JJ, Aparicio O, Jennings EG, Simon I, Zeitlinger J, Schreiber J, Hannett N, Kanin E (2000) Genome-wide location and function of DNA binding proteins. Science 290(5500):2306–2309

Ritala A, Häkkinen ST, Schillberg S (2014) Molecular pharming in plants and plant cell cultures: a great future ahead? Pharm Bioprocess 2(3):223–226

Rocca-Serra P, Brazma A, Parkinson H, Sarkans U, Shojatalab M, Contrino S, Vilo J, Abeygunawardena N, Mukherjee G, Holloway E, Kapushesky M, Kemmeren P, Lara GG, Oezcimen A, Sansone SA (2003) ArrayExpress: a public database of gene expression data at EBI. C R Biol 326(10–11):1075–1078

Rose PW, Bi C, Bluhm WF, Christie CH, Dimitropoulos D, Dutta S, Green RK, Goodsell DS, Prlić A, Quesada M (2013) The RCSB protein data bank: new resources for research and education. Nucleic Acids Res 41(D1):D475–D482

Rusyn I, Daston GP (2010) Computational toxicology: realizing the promise of the toxicity testing in the 21st century. Environ Health Perspect 118(8):1047–1050

Saito K (2013) Phytochemical genomics—a new trend. Curr Opin Plant Biol 16(3):373–380

Samwald M, Dumontier M, Zhao J, Luciano JS, Marshall MS, Cheung K (2010) Integrating findings of traditional medicine with modern pharmaceutical research: the potential role of linked open data. Chinese Med 5:43

Sarkar IN (2000) Biodiversity informatics: the emergence of a field. BMC Bioinforma 10(Suppl 14):S1

Saxena M, Saxena J, Nema R, Singh D, Gupta A (2013) Phytochemistry of medicinal plants. J Pharmacogn Phytochem 1(6):168–182

Sayers EW, Barrett T, Benson DA, Bolton E, Bryant SH, Canese K, Chetvernin V, Church DM, Dicuccio M, Federhen S, Feolo M, Geer LY, Helmberg W, Kapustin Y, Landsman D, Lipman DJ, Lu Z, Madden TL, Madej T, Maglott DR, Marchler-Bauer A, Miller V, Mizrachi I, Ostell J, Panchenko A, Pruitt KD, Schuler GD, Sequeira E, Sherry ST, Shumway M, Sirotkin K, Slotta D, Souvorov A, Starchenko G, Tatusova TA, Wagner L, Wang Y, John Wilbur W, Yaschenko E, Ye J (2010) Database resources of the national center for biotechnology information. Nucleic Acids Res 38(D1):D5–D16

Schena M, Shalon D, Davis RW, Brown PO (1995) Quantitative monitoring of gene expression patterns with a complementary DNA microarray. Science 270(5235):467

Schläpfer P, Zhang P, Wang C, Kim T, Banf M, Chae L, Dreher K, Chavali AK, Nilo-Poyanco R, Bernard T, Kahn D, Rhee SY (2017) Genome-wide prediction of metabolic enzymes, pathways, and gene clusters in plants. Plant Physiol 173(4):2041–2059

Sharma V, Sarkar IN (2012) Bioinformatics opportunities for identification and study of medicinal plants. Brief Bioinform 14(2):238–250

Singh A, Kumar N (2013) A review on DNA microarray technology. Int J Curr Res Rev 5(22):1

Smith CA, O'Maille G, Want EJ, Qin C, Trauger SA, Brandon TR, Custodio DE, Abagyan R, Siuzdak G (2005) METLIN: a metabolite mass spectral database. Ther Drug Monit 27(6):747–751

Strohl WR (2000) The role of natural products in a modern drug discovery program. Drug Discov Today 5(2):39–41

Tanabe L, Scherf U, Smith LH, Lee JK, Michaels GS, Hunter L, Weinstein JN (1999) MedMiner: an internet tool for filtering and organizing biomedical information. BioTechniques 27:1210–1217

Tatusova T, DiCuccio M, Badretdin A, Chetvernin V, Nawrocki EP, Zaslavsky L, Lomsadze A, Pruitt KD, Borodovsky M, Ostell J (2016) NCBI prokaryotic genome annotation pipeline. Nucleic Acids Res 44(14):6614–6624

Ueno S, Moriguchi Y, Uchiyama K, Ujino-Ihara T, Futamura N, Sakurai T, Shinohara K, Tsumura Y (2012) A second generation framework for the analysis of microsatellites in expressed sequence tags and the development of EST-SSR markers for a conifer, *Cryptomeria japonica*. BMC Genomics 13(1):1

UniProt C (2014) Activities at the universal protein resource (UniProt). Nucleic Acids Res 42(D1):D191–D198

van Dijk EL, Auger H, Jaszczyszyn Y, Thermes C (2014) Ten years of next-generation sequencing technology. Trends Genet 30(9):418–426

Valerio LG Jr, Arvidson KB, Busta E, Minnier BL, Kruhlak NL, Benz RD (2010) Testing computational toxicology models with phytochemicals. Mol Nutr Food Res 54(2):186–194

Varshney RK, Graner A, Sorrells ME (2005) Genic microsatellite markers in plants: features and applications. Trends Biotechnol 23(1):48–55

Wagner JC, Rogers JE, Baud RH, Scherrer JR (1998) Natural language generation of surgical procedures. Stud Health Technol Inform 52(Pt 1):591–595

Wang JF, Zhou H, Han LY, Chen X, Chen YZ, Cao ZW (2005) Traditional Chinese medicine information database. Clin Pharmacol Ther 78(1):92–93

Wang Y, Wang X, Cheng Y (2006) A computational approach to botanical drug design by modeling quantitative composition-activity relationship. Chem Biol Drug Des 68(3):166–172

Wetie AGN, Sokolowska I, Woods AG, Roy U, Deinhardt K, Darie CC (2014) Protein–protein interactions: switch from classical methods to proteomics and bioinformatics-based approaches. Cell Mol Life Sci 71(2):205–228

Wheeler DL, Barrett T, Benson DA, Bryant SH, Canese K, Chetvernin V, Church DM, DiCuccio M, Edgar R, Federhen S (2007) Database resources of the national center for biotechnology information. Nucleic Acids Res 35(Suppl 1):D5–D12

Wise RP, Caldo RA, Hong L, Shen L, Cannon E, Dickerson JA (2007) BarleyBase/PLEXdb. Methods Mol Biol 406:347–363

Wolfender J-L, Rudaz S, Hae Choi Y, Kyong Kim H (2013) Plant metabolomics: from holistic data to relevant biomarkers. Curr Med Chem 20(8):1056–1090

Wootton JC (2002) "Development of HerbMed": an interactive. Ethnomed Drug Discov 1:55

Wu KM, Farrelly J, Birnkrant D, Chen S, Dou J, Atrakchi A, Bigger A, Chen C, Chen Z, Freed L, Ghantous H, Goheer A, Hausner E, Osterberg R, Rhee H, Zhang K (2004) Regulatory toxicology perspectives on the development of botanical drug products in the United States. Am J Ther 11(3):213–217

Xu Y, Lu Y, Xie C, Gao S, Wan J, Prasanna BM (2012) Whole-genome strategies for marker-assisted plant breeding. Mol Breed 29(4):833–854

Xue R, Fang Z, Zhang M, Yi Z, Wen C, Shi T (2013) TCMID: traditional Chinese medicine integrative database for herb molecular mechanism analysis. Nucleic Acids Res 41(D1):D1089–D1095

Yang C, Hasselgren CH, Boyer S, Arvidson K, Aveston S, Dierkes P, Benigni R, Benz RD, Contrera J, Kruhlak NL, Matthews EJ, Han X, Jaworska J, Kemper RA, Rathman JF, Richard AM (2008) Understanding genetic toxicity through data mining: the process of building knowledge by integrating multiple genetic toxicity databases. Toxicol Mech Methods 18(2–3):277–295

Yang J, Roy A, Zhang Y (2013) Protein–ligand binding site recognition using complementary binding-specific substructure comparison and sequence profile alignment. Bioinformatics 29(20):2588–2595

Yang D, Du X, Yang Z, Liang Z, Guo Z, Liu Y (2014) Transcriptomics, proteomics, and metabolomics to reveal mechanisms underlying plant secondary metabolism. Eng Life Sci 14(5):456–466

Ye H, Ye L, Kang H, Zhang D, Tao L, Tang K, Liu X, Zhu R, Liu Q, Chen YZ, Li Y, Cao Z (2011) HIT: linking herbal active ingredients to targets. Nucleic Acids Res 39(D1):D1055–D1059

Youns M, Hoheisel JD, Efferth T (2010) Toxicogenomics for the prediction of toxicity related to herbs from traditional Chinese medicine. Planta Med 76:2019–2025

Zeng X, Zhang P, He W, Qin C, Chen S, Tao L, Wang Y, Tan Y, Gao D, Wang B, Chen Z, Chen W, Jiang YY, Chen YZ (2018) NPASS: natural product activity and species source database for natural product research, discovery and tool development. Nucleic Acids Res 46(D1):D1217–D1222

Zhang L, Berleant D, Ding J, Cao T, Wurtele ES (2009) PathBinder–text empirics and automatic extraction of biomolecular interactions. BMC Bioinforma 10(11):1

Zhang RZ, Yu SJ, Bai H, Ning K (2017) TCM-mesh: the database and analytical system for network pharmacology analysis for TCM preparations. Sci Rep 7(1):2821

Zhao XP, Fan XH, Yu J, Cheng YY (2004) A method for predicting activity of traditional Chinese medicine based on quantitative composition-activity relationship of neural network model. Zhongguo Zhong Yao Za Zhi 29(11):1082–1085

Chapter 9
Experimental Approaches for Genome Sequencing

Mohd Sayeed Akhtar, Ibrahim A. Alaraidh, and Khalid Rehman Hakeem

Contents

9.1   Introduction... 159
9.2   Genome Sequencing Approaches.. 160
     9.2.1   BAC-to-BAC Approach... 160
     9.2.2   Shotgun Approach... 161
     9.2.3   Other Sequencing Approaches...................................... 162
9.3   Conclusions and Future Prospects.. 164
References... 164

9.1   Introduction

Genome sequencing is developing rapidly as a revolutionizing field due to advances in DNA sequencing technologies and started the new era in the field of molecular biology (Kelley and Salzberg 2010; Ruperao et al. 2014; Khan et al. 2016; Visendi et al. 2016). The scientist working in this arena has gained the popularity by manipulating the DNA molecules for the study of genes and their harness toward the development sparking a new revolution in biological investigations (Fuller et al. 2006; Hsu et al. 2014). These recent advances in genome sequencing served as an important tool in basic and translational research, drug development, and clinical

M. S. Akhtar (✉)
Department of Botany, Gandhi Faiz-e-Aam College, Shahjahanpur, Uttar Pradesh, India

I. A. Alaraidh
Botany and Microbiology Department, King Saud University, Science College, Riyadh, Saudi Arabia

K. R. Hakeem
Department of Biological Sciences, King Abdulaziz University, Jeddah, Saudi Arabia

Princess Dr. Najla Bint Saud Al-Saud Center for Excellence Research in Biotechnology, King Abdulaziz University, Jeddah, Saudi Arabia

© Springer Nature Switzerland AG 2019   159
K. R. Hakeem et al. (eds.), *Essentials of Bioinformatics, Volume III*,
https://doi.org/10.1007/978-3-030-19318-8_9

trials (Fontana et al. 2012; Readhead and Dudley 2013; Roychowdhury and Chinnaiyan 2016). Nowadays, the sequencing cost and a high-throughput data generation are not limiting factors due to development of modern sophisticated technologies, but a big core facility is still needed to operate the procedures of genome sequencing (Buermans and den Dunnen 2014; Head et al. 2014). Since, the genome analysis is increasingly used to address various problems related to the genotyping, diagnosis, environmental and microbiome profiling, and mutation and evolutionary studies. The number of challenges in genome analysis is associated with sequencing methods. There has been very fast development in genome sequencing. A good example was the completion of human genome project in record time. The human body has about 100 trillion cells. Inside each cell is the nucleus that contains the genome (46 human chromosomes), which governs human development (Kothekar and Nandi 2007). Similarly, the tomato genome identified an esterase responsible for differences in volatile ester content in different tomato species (Goulet et al. 2012). In general, the chromosomes comprise millions of copies of the four-letter genetic code—the DNA bases (A, C, G, and T) which are arranged into genes and noncoding sections (Akhtar et al. 2017). Finding the order or sequence of these four letters is the goal in genomics. The entire human genome is made up of about 3.5 billion bases. To read the DNA sequence, the chromosomes are cut into tiny pieces and read individually. When all the segments have been read, they assembled in the correct order. The properties of a biological system are studied through the expression of many genes simultaneously. Simple interpretation strategies are useful. A typical example is protein p53, an early component in cells which respond to DNA damage (Goeman et al. 2017). Thus, the aim of this chapter is to provide the in-depth knowledge of various experimental approaches used for sequencing of the genome.

9.2 Genome Sequencing Approaches

There are fundamentally two ways to sequence the genome, namely BAC (bacterial artificial chromosome)-to-BAC approach and shotgun approach.

9.2.1 BAC-to-BAC Approach

BAC-to-BAC approach is also referred to as the map-based approach. It was first employed in human genome studies during the late 1980s and continues its expansion till date. The BAC-to-BAC approach first creates a crude physical map of the whole genome (Becker 1998; Bolger et al. 2014). Constructing a map requires cutting the chromosomes into pieces and figuring out the order of the big chunks of DNA before sequencing all the fragments. The BAC sequences were individually assembled and arranged according to the physical map, creating a very high-quality genome sequence (Fig. 9.1a).

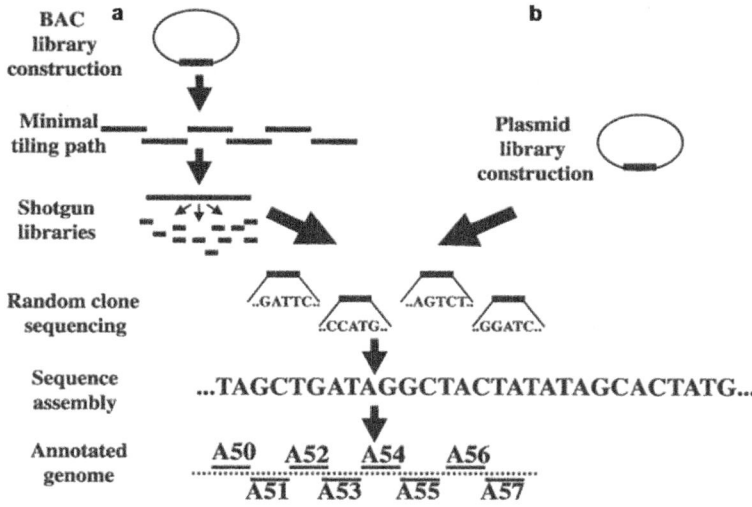

Fig. 9.1 Whole genome sequencing; (**a**) BAC-to-BAC approach; (**b**) Shotgun approach

Several copies of the genome are randomly cut into pieces of about 150,000 base pairs (bp), and each of these fragments is inserted into a BAC. A BAC is an artificial piece of DNA that can replicate inside the bacterial cell. The whole collection of BACs containing the entire human genome is called a BAC library. Each BAC is like a book in the library that can be accessed and copied. Each piece is finger-printed to give a unique identification tag that determines the order of the fragments. Fingerprinting involves cutting each BAC fragment with a single enzyme and find-ing common sequence landmarks in overlapping fragments that determine the loca-tion of each BAC along the chromosome. Each BAC is then broken randomly into pieces of 1500 bp and is placed in another artificial piece of DNA called M13. The collection is known as M13 library. All the M13 libraries are sequenced. A 500 bp from one end of the fragment is sequenced to generate millions of sequences. Compute algorithms assemble millions of sequenced fragments into a continuous stretch of the chromosome. These sequences are fed into a computer program called PHRAP, which looks for common sequences that join two fragments.

9.2.2   Shotgun Approach

It is a speedy approach to genome sequencing, which may enable the researchers to complete their job in a short time. Venter (1996) developed the shotgun approach at The Institute for Genomic Research (TIGR). The approach of sequencing bypasses the need for a physical map and goes straight into the job of decoding (Fig. 9.1b). This is the main reason for this speedy technique. Multiple copies of the genome are randomly shredded into pieces of 2000 bp by squeezing the DNA through a pressur-ized syringe. This is done a second time to generate pieces of 10,000 bp long.

Each 2000 and 10,000 bp fragment is inserted into a plasmid, which is a piece of DNA that replicates in bacteria. The two collections of plasmids containing 2000 and 10,000 bp chunks of human DNA are known as plasmid libraries. Both 2000–10,000 bp plasmid libraries are sequenced. A 500 bp from each end of a fragment is decoded to generate millions of sequences. Sequencing both ends is critical for assembling the entire chromosome; computer algorithms assemble millions of sequenced fragments into a continuous stretch of the chromosome.

9.2.3   Other Sequencing Approaches

9.2.3.1   Large-Scale Approach

It includes hybridization and sequencing approaches. Hybridization has evolved from early membrane-based radioactive detection embodiments to parallel quantitative methods using fluorescence detection (Lee 2007).

9.2.3.2   cDNA Microarray Detectors

It is a very sensitive technique. It requires only 2–5 nl of DNA solution coating with poly L-lysine and aminocialines. The cDNA libraries provide a flexible sequence probe. The choice of fluoroprobe is important. The biological samples (or their cDNA derivatives) are hybridized to the range and are referred to as the target. Labeling with fluorescent dyes with different excitation and emission characteristics allows the simultaneous hybridization of two contrasting targets on a single array (Aharoni and Vorst 2001; Campos-De Quiroz 2002). Microarrays can be based on cDNA molecules, and their basic features are represented in tabular form (Table 9.1).

9.2.3.3   PCR Method

PCR is used to amplify the single copy or copies of a DNA segments across several orders of magnitude to generate millions of copies of a desired DNA sequences. It is an easy, cheap, and reliable technique to replicate a focused segment of DNA, and most widely used in molecular biology for biomedical research, criminal forensics, and molecular archaeology. Now, it is commonly used in clinical and research

Table 9.1 Some common features of cDNA microarray

Features	cDNA microarray
Array preparation	Direct or indirect spotting
Target	cDNA
Target labeling	Cy3-dCTP and Cy5-dCTP incorporation through reverse transcription
Type of hybridization	DNA-DNA

laboratories for a broad variety of applications including DNA cloning and manipulation, gene mutagenesis, construction of DNA-based phylogenies, or functional analysis of genes; diagnosis and monitoring of hereditary diseases; amplification of ancient DNA; analysis of genetic fingerprints; and detection of pathogens in nucleic acid tests for the diagnosis of infectious diseases. The majority of PCR methods rely on thermal cycling, which involves exposing the reactants to cycles of repeated heating and cooling, permitting different temperature-dependent reactions specifically DNA melting and enzyme-driven DNA replication to quickly proceed many times in sequence. Primers contain sequences complementary to the target region, along with a DNA polymerase to enable selected and repeated amplifications. In PCR, the choice of template is important. As the PCR progresses, the DNA generated is itself used as a template for replication, setting in motion a chain reaction in which the original DNA template is exponentially amplified. Almost all PCR applications employ a heat-stable DNA polymerase, such as Taq polymerase. If heat-susceptible DNA polymerase is used, it will denature every cycle at the denaturation step. Before the use of Taq polymerase, DNA polymerase had to be manually added every cycle, which was a tedious and costly process. This DNA polymerase enzymatically assembles a new DNA strand from free nucleotides by using single-stranded DNA as a template and DNA oligonucleotides to initiate DNA synthesis. In the first step, the two strands of the DNA double helix are physically separated at a high temperature in a process called DNA melting. In the second step, the temperature is lowered and the two DNA strands become templates for DNA polymerase to selectively amplify the target DNA. Selectivity of PCR results from the use of primers that are complementary to sequence around the DNA region targeted for amplification under specific thermal cycling conditions. PCR has an enormous impact in both basic and diagnostic aspects of molecular science because it produces large amounts of a specific DNA fragments from small amounts of a complex template. PCR represents a form of in vitro cloning that can generate or modify the DNA fragments of definite length and sequence in a simple automated reaction. In addition, PCR plays a critical role in the identification of medically important sequences as well as an important diagnostic one in their detection.

9.2.3.4   Image Analysis

Intensity evaluation (100–400 μ^2 pixels) allows 50–200 sampling at each spot. Data normalization is done. Statistical analysis is important. Assay reliability has to be tested.

9.2.3.5   Functional Proteomics

Phene is the functional protein contained in the total protein data. Phene is to "phenotype" as a gene is to "genotype." In multicellular organisms, the set of proteins would differ from cell type to cell type. Proteins normally undergo large-scale modifications. Proteome analysis is concerned with biochemical changes like posttranslational modifications, phosphorylation, etc. In general, the phosphorylation is a reversible enzymatic reaction and plays an important role in various cellular processes, viz., division, function of target proteins, immunity metabolism, membrane transport, and organelle trafficking (Bolger et al. 2014). It can activate and inhibit enzyme activity through allosteric conformational changes, facilitate the recognition of other proteins, promote protein-protein association or dissociation, and also induce order to disorder transition.

9.3 Conclusions and Future Prospects

Genome sequencing has a major impact on molecular biological research and improvement in the comparatively small period of time. Although a rapid development has been observed in the preparation of library in the past decades with the performances of some small genome studies, still the breakthrough researches are expected. Therefore, in the future, more studies are needed on the deep phenotyping platforms to overcome the issues and the elucidation of mechanisms to complete enormity of the available data.

Acknowledgments The authors (Mohd Sayeed Akhtar and Ibrahim A. Alaraidh) are highly grateful to the Department of Botany, Gandhi Faiz-e-Aam College, Shahajahanpur, U.P., India, and the Botany and Microbiology Department, Science College, King Saud University, Riyadh, Kingdom of Saudi Arabia.

References

Aharoni A, Vorst O (2001) DNA microarrays for functional plant genomics. Plant Mol Biol 48:99–118

Akhtar MS, Swamy MK, Alaraidh IA, Panwar J (2017) Genomic data resources and data mining. In: Hakeem KR, Malik A, Sukan FV, Ozturk M (eds) Plant bioinformatics. Springer International Publishing, Cham, pp 267–278

Becker S (1998) Three-dimensional structure of the Stat3β homodimer bound to DNA. Nature 394:145–151

Bolger ME, Weisshaar B, Scholz U, Stein N, Usadel B, Mayer KFX (2014) Plant genome sequencing applications for crop improvement. Curr Opin Biotechnol 26:31–37

Buermans HPJ, den Dunnen JT (2014) Next generation sequencing technology: advances and applications. Biochim Biophys Acta 1842:1932–1941

Campos-De Quiroz H (2002) Plant genomics: an overview. Biol Res 35:385–399

Fontana JM, Alexander E, Salvatore M (2012) Translational research in infectious disease: current paradigms and challenges ahead. Transl Res 159:430–453

Fuller DN, Gemmen GJ, Rickgauer JP, Dupont A, Millin R, Recouvreux P, Smith DE (2006) A general method for manipulating DNA sequences from any organism with optical tweezers. Nucleic Acids Res 34:e15. https://doi.org/10.1093/nar/gnj016

Goeman F, Strano S, Blandino G (2017) MicroRNAs as key effectors in the p53 network. Int Rev Cell Mol Biol 333:51–90

Goulet C, Mageroy MH, Lam NB, Floystad A, Tieman DM, Klee HJ (2012) Role of an esterase in flavor volatile variation within the tomato clade. Proc Natl Acad Sci U S A 109:19009–19014

Head SR, Komori HK, LaMere SA, Whisenant T, Nieuwerburgh FV, Salomon DR, Ordoukhanian P (2014) Library construction for next-generation sequencing: overviews and challenges. Biotechniques 56:61. https://doi.org/10.2144/000114133

Hsu PD, Lander ES, Zhang F (2014) Development and applications of CRISPR-Cas9 for genome engineering. Cell 157:1262–1278

Kelley DR, Salzberg SL (2010) Detection and correction of false segmental duplications caused by genome mis-assembly. Genome Biol 11:R28

Khan S, Wajid Ullah M, Siddique R, Nabi G, Manan S, Yousaf M, Hou H (2016) Role of recombinant DNA technology to improve life. Int J Genomics 2016:2405954. https://doi.org/10.1155/2016/2405954

Kothekar V, Nandi T (2007) An introduction to bioinformatics. Duckworth Press, BioScience Publishers, Delhi

Lee MLT (2007) Analysis of microarray gene expression data. Kluwer Academic Publisher, Boston

Readhead B, Dudley J (2013) Translational bioinformatics approaches to drug development. Adv Wound Care 2:470–489. https://doi.org/10.1089/wound.2012.0422

Roychowdhury S, Chinnaiyan AM (2016) Translating cancer genomes and transcriptomes for precision oncology. CA Cancer J Clin 66:75–88

Ruperao P, Chan C-KK, Azam S, Karafiátová M, Hayashi S, Čížková J, Saxena RK, Šimková H, Song C, Vrána J, Chitikineni A, Visendi P, Gaur PM, Millán T, Singh KB, Tara B, Wang J, Batley J, Doležel J, Varshney RK, Edwards D (2014) A chromosomal genomics approach to assess and validate the desi and kabuli draft chickpea genome assemblies. Plant Biotechnol J 12:778–786

Venter JC, Smith HO, Leed H (1996) A new strategy for genome sequencing. Nature 381:364–366

Visendi P, Berkman PJ, Hayashi S, Golicz AA, Bayer PE, Ruperao P, Hurgobin B, Montenegro J, Chan CKK, Staňková H, Batley J, Šimková H, Doležel J, Edwards D (2016) An efficient approach to BAC based assembly of complex genomes. Plant Methods 12:2. https://doi.org/10.1186/s13007-016-0107-9

Chapter 10
Phylogenetic Trees: Applications, Construction, and Assessment

Surekha Challa and Nageswara Rao Reddy Neelapu

Contents

10.1		Introduction and Applications of Phylogeny	167
	10.1.1	Affiliating Taxonomy to an Organism	168
	10.1.2	Studying Reproductive Biology in Lower Organisms	169
	10.1.3	Assessing the Process of Cryptic Speciation in a Species	170
	10.1.4	Studying the Evolution of Proteins or Gene Families	172
	10.1.5	Classifying Proteins or Genes into Families	174
	10.1.6	Understanding the History of Life	175
	10.1.7	Estimating the Time of Divergence Using Molecular Clock	175
	10.1.8	Evolution of Pathogen	178
10.2		Construction of Phylogenetic Trees	181
	10.2.1	Data	181
	10.2.2	Tree-Constructing Methods	182
	10.2.3	Phylogeny Program Packages	183
10.3		Methods to Assess the Confidence of Phylogenetic Tree	183
	10.3.1	Sampling Methods	183
	10.3.2	Statistical Methods	188
10.4		Conclusion	189
References			189

10.1 Introduction and Applications of Phylogeny

Study of relationships among individuals or groups of organisms or species or populations is called phylogeny. The relationships among the individuals are estimated or assessed based on the evolutionary signals present in the genetic material of any organism. The evolutionary signals or footprints among these individuals or entities are used to construct the evolutionary history. The evolutionary history based on the evolutionary signals can be modeled or represented in the form of graphical

S. Challa · N. R. R. Neelapu (✉)
Department of Biochemistry and Bioinformatics, Institute of Science, Gandhi
Institute of Technology and Management (GITAM) (Deemed to be University),
Visakhapatnam, Andhra Pradesh, India

© Springer Nature Switzerland AG 2019   167
K. R. Hakeem et al. (eds.), *Essentials of Bioinformatics, Volume III*,
https://doi.org/10.1007/978-3-030-19318-8_10

representation or tree, which is known as phylogenetic tree. Phylogenetics is an ever-evolving field that promises to give more insights into understanding biodiversity, evolution, ecology, and genomes. Phylogenetics has several applications like affiliating taxonomy to an organism, studying reproductive biology in lower organisms, assessing the process of cryptic speciation in a species, understanding the history of life, resolving controversial history of life, reconstructing the paths of infection in an epidemiology to understand the evolution of pathogen, classifying proteins or genes into families, and many more.

10.1.1 Affiliating Taxonomy to an Organism

Every living organism which is known or identified till date should be classified and affiliated to a taxonomic group. When the taxonomy of the species identified is not known, it is left as an orphan or classified into a special group. The traditional approach for identification of an organism includes studies based on microscopy, morphology, biochemical tests, physiological tests, fruit bodies, mating behavior experiments, and others. The drawbacks associated with the traditional approach are time consuming and of low to moderate in precision. In these cases, phylogeny can be used to affiliate taxonomy to a taxa or an organism.

Phylogeny has been proposed and widely accepted to affiliate taxonomy for a species. Several reports were there on entomopathogenic fungi (Neelapu et al. 2009), *Echinococcus* (Thompson 2008), catfishes (Teugels 1996), *Borrelia burgdorferi* (Margos et al. 2011), *Trichinella* (Pozio et al. 2009), and many more. This case study provides with details that how phylogeny can be used to affiliate taxonomy for entomopathogenic fungi (Neelapu et al. 2009). When the taxonomy of the species is not known, it is left as an orphan or classified into a special group. The fungi which are not classified into any fungal divisions such as *Ascomycota*, *Zygomycota*, and *Basidiomycota* were classified into a special group known as Deuteromycota. Neelapu et al. (2009) studied phylogeny of mitosporic or asexual or conidiogenous entomopathogenic fungi of Deuteromycota belonging to the genera *Beauveria*, *Nomuraea*, *Metarhizium*, *Paecilomyces*, and *Lecanicillium*. One hundred forty-seven fungal entries covering 94 species related to *Ascomycota*, *Zygomycota*, and *Basidiomycota* were analyzed. The partial amino acid sequences of the β-tubulin gene were aligned using AlnExplorer of MEGA ver. 3.014. The statistical procedures minimum evolution (ME), maximum parsimony (MP), and neighbor joining (NJ) of MEGA ver. 3.014; maximum likelihood of PAUP ver. 4b; Bayesian inference of MrBayes ver. 3.04b10; and Metropolis-coupled Markov chain Monte Carlo (MCMCMC) were used to construct phylogenetic tree. "Phylogenetic analysis placed all the asexual entomopathogenic fungal species analyzed in the family *Clavicipitaceae* of the order *Hypocreales* of *Ascomycota*" (Fig. 10.1). Thus, whenever the identity of the organism is in crisis, phylogeny can be used to affiliate the organism to the known traditional taxonomic group.

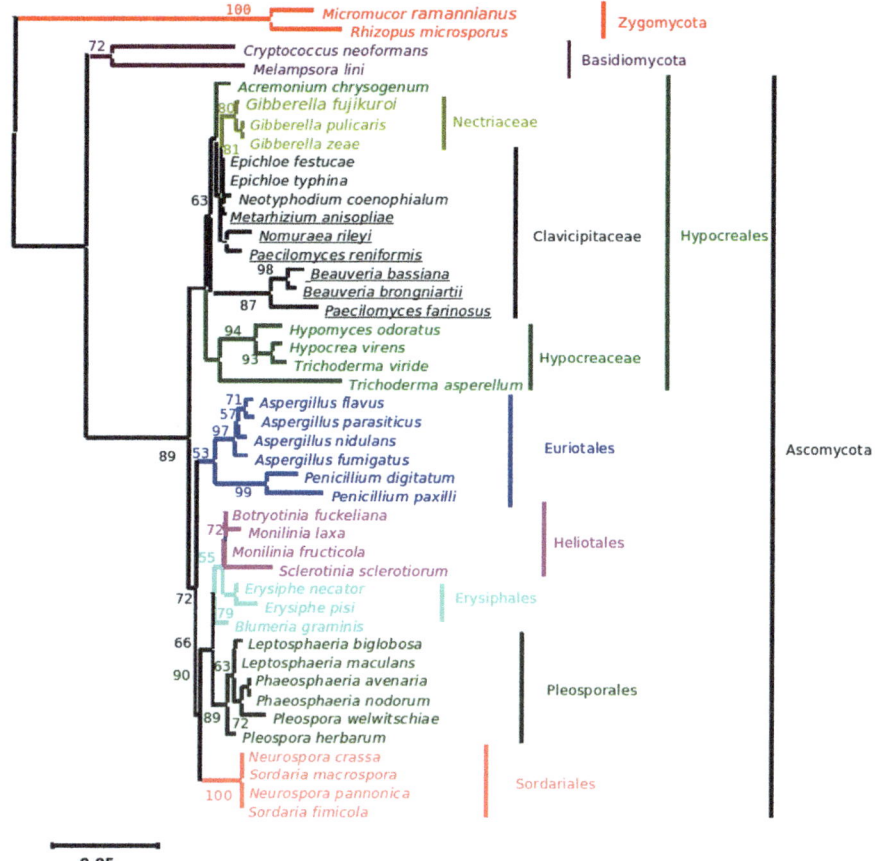

Fig. 10.1 The phylogenetic affiliation of the asexual entomopathogenic *Beauveria* spp., *Nomuraea* spp., *Metarhizium* spp., and *Paecilomyces* spp. (Source: Neelapu et al. 2009)

10.1.2   Studying Reproductive Biology in Lower Organisms

Understanding the reproductive biology in lower organisms where sexual organs are not observed is a challenge. Genetic tests based on phylogenetic concordance and gene genealogies offer an indirect means of identifying recombination. When phylogeny is applied, different genes show different genealogies within a species due to recombination. Therefore, phylogenetic trees generated from the data show phylogenetic concordance among the multiple gene genealogies in recombining species, whereas non-phylogenetic concordance among the multiple gene genealogies in a clonal species (Fig. 10.2).

The reproductive biology in *Beauveria bassiana* (Neelapu 2007; Devi et al. 2006) and *Nomuraea rileyi* (Neelapu 2007; Devi et al. 2007) was studied. Devi et al. (2006)

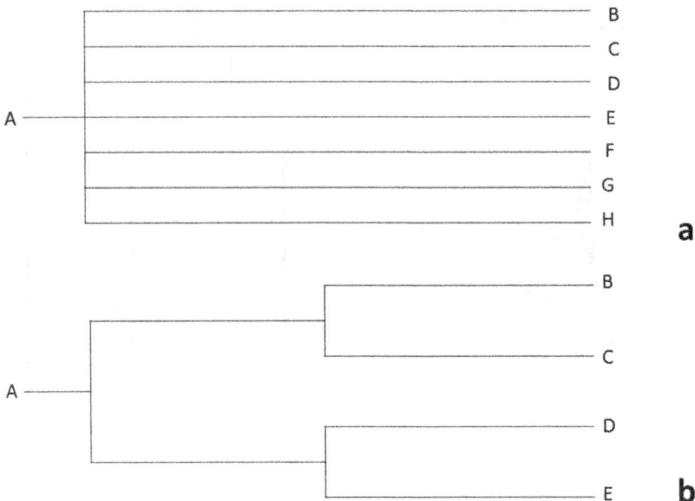

Fig. 10.2 Phylogenetic concordance and gene genealogies: (**a**) clonal species (**b**) recombining species

applied indirect means of genetic tests which are based on phylogenetic concordance of gene genealogies to identify reproductive biology (recombination or clonal) in a localized epizootic population of entomopathogenic fungi *B. bassiana*. Nucleotide sequence data of different allelic forms of three genes (large and small subunits of mitochondrial ribosomal RNA (mt rRNA) and β-tubulin) were evaluated to assess phylogenetic concordance among the multiple gene genealogies. Lack of phylogenetic concordance among three gene genealogies in the epizootic of *B. bassiana* indicates prevalence of recombination within the clonal structure of the population (Fig. 10.3). Thus, whenever the mating tests cannot be applied in lower organisms like bacteria and fungi where sexual organs are not observed, phylogenetic concordance among multiple gene genealogies can be used for understanding the reproductive biology.

10.1.3 Assessing the Process of Cryptic Speciation in a Species

Entomopathogenic fungi of Deuteromycota belonging to the genera *Beauveria*, *Nomuraea*, *Metarhizium*, and *Paecilomyces* are recognized as a "species complex" comprising of genetically diverse lineages. Devi et al. (2006) used amplified fragment length polymorphism (AFLP) and single-stranded confirmation polymorphism (SSCP) data of worldwide population and generated unweighted pair group method with arithmetic mean (UPGMA) tree. The worldwide sample of *B. bassiana* isolates represented cryptic phylogenetic species (Fig. 10.4). Literature reports the use of powerful approach—genealogical concordance phylogenetic species recognition (GCPSR)—to uncover cryptic speciation. "GCPSR detects genetically isolated

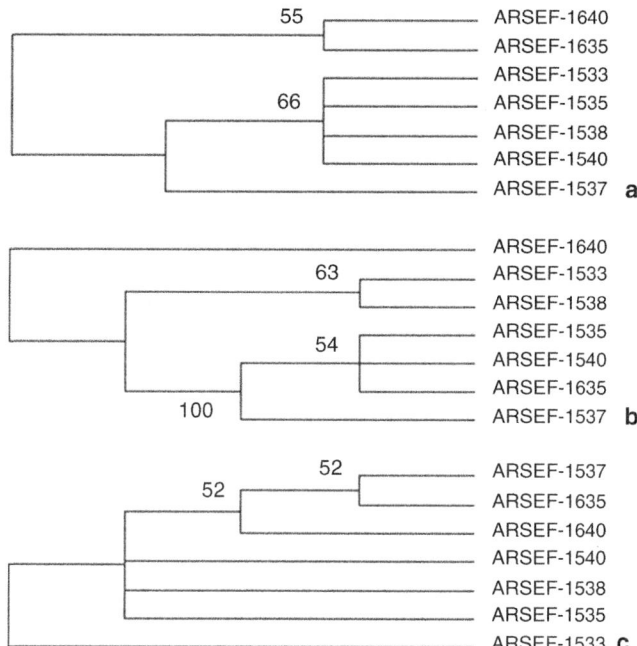

Fig. 10.3 Maximum parsimony tree generated from the sequences of (**a**) partial sequence of β-tubulin gene, (**b**) large subunit of mt rRNA gene, and (**c**) small subunit of mt rRNA genes derived from the isolates of an epizootic *B. bassiana* population from Burgenland, Austria. The tree topology of each species tree indicates the presence of recombination and cryptic speciation. (Source: Devi et al. 2006)

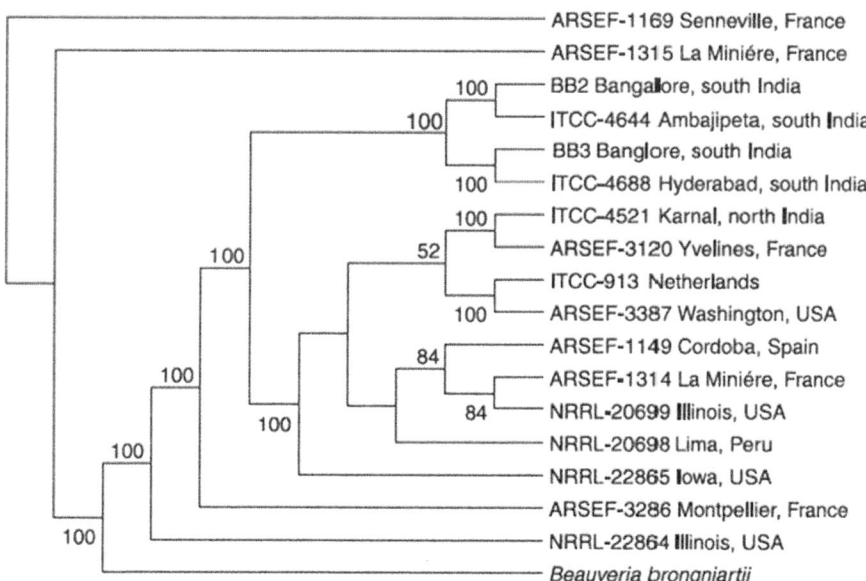

Fig. 10.4 Phylogenetic tree derived and generated from SSCP data of three genes: β-tubulin gene, and large and small subunits of mt rRNA genes of a sample of isolates of *B. bassiana* of worldwide distribution, representing cryptic phylogenetic species (Neelapu et al. 2009)

groups from a number of different loci by comparing the gene trees. Different genes have different genealogies within a species establishing gene flow delimiting species by identifying the unshared polymorphisms, and thus branches that are incompatible, with all genealogies at all loci. Thus, branches that are incompatible with all genealogies at all loci represent different species" (Neelapu et al. 2009).

Neelapu et al. (2009) used GCPSR to uncover cryptic speciation in *B. bassiana*. Epizootic population of *B. bassiana* from Burgenland, Austria, are sequenced for partial sequences of the three genes, β-tubulin gene and large and small subunit of rRNA genes of mitochondria, and were aligned using AlnExplorer of MEGA ver. 3.1. A consensus maximum parsimony tree was generated using PAUP ver. 4.0. "The tree topology of each species tree indicates the presence of cryptic speciation. Incongruity of gene genealogies within a given group indicates gene flow and delimits a species. As the approach detects reproductive isolation, the resulting groups also fulfill the criteria of a biological species" (Fig. 10.3).

10.1.4 Studying the Evolution of Proteins or Gene Families

Phylogeny is used in establishing the origin and evolutionary pattern of a gene of particular species with respect to the other species. Similar set of genes are required for studying or understanding the phylogeny. The genes, which are similar in their structure or function, are known as homologous sequences. If the genes are similar in function but are from different organisms, then they are believed to be orthologous sequences. If the genes are from the same organism, then they are known as paralogous sequences. It is believed that orthologous sequences are due to speciation from a common ancestor, whereas paralogous sequences are due to duplication.

Though there are many reports on the evolution of proteins or gene families, we would like to throw some light on evolution of globin and V-PPases (Hardison 2012; Suneetha et al. 2016). Globin genes diverged to form hemoglobin (oxygen transport in blood), myoglobin (oxygen metabolism in muscle), cytoglobin (oxygen donator during synthesis and cross-linking of collagen or acting as a protector of the free radicals formed in the fibrosis process), and neuroglobin (acts as an oxygen reservoir releasing oxygen in stressful situations, such as hypoxia). So, the plausible explanation for gene evolution can be duplication of the existing gene like globin followed by divergence in function as described above for hemoglobin, myoglobin, cytoglobin, and neuroglobin (Figs. 10.3 and 10.4) (Hardison 2012). The best example for both orthologous and paralogous sequence is globin genes. α-Globin and β-globin genes found in different species are orthologous genes (Fig. 10.5), whereas the α, β, γ, and δ globin genes due to duplication in the same organism are paralogous genes (Fig. 10.6) (Hardison 2012; Opazo et al. 2008).

V-PPase is a heat-stable single polypeptide, coexisting along with V-ATPase on the plant vacuolar membrane in plants, algae, photosynthetic bacteria, protozoa, and archaebacteria (Rea et al. 1992; Maeshima 2000). V-PPase uses ATP and inorganic pyrophosphate (PPi), respectively, as energy sources for generating an electrochemical gradient of protons across the tonoplast. This facilitates the functioning of the Na^+/H^+

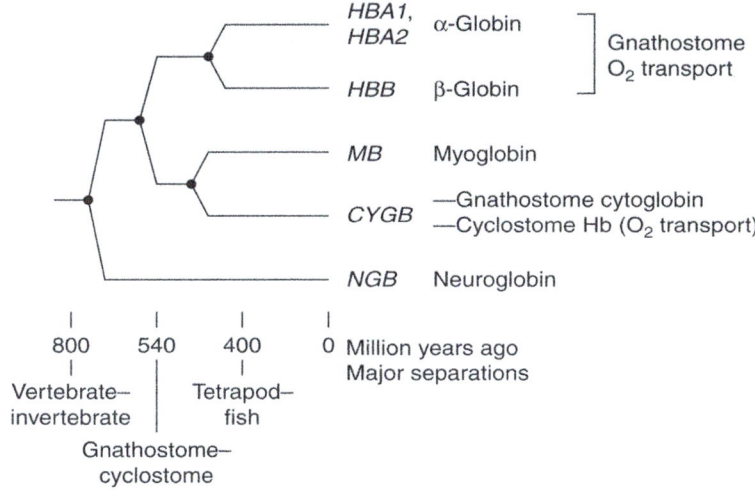

Fig. 10.5 Phylogenetic tree showing duplication and divergence of globin genes, an example for evolution of vertebrate globin genes. (Source: Hardison 2012)

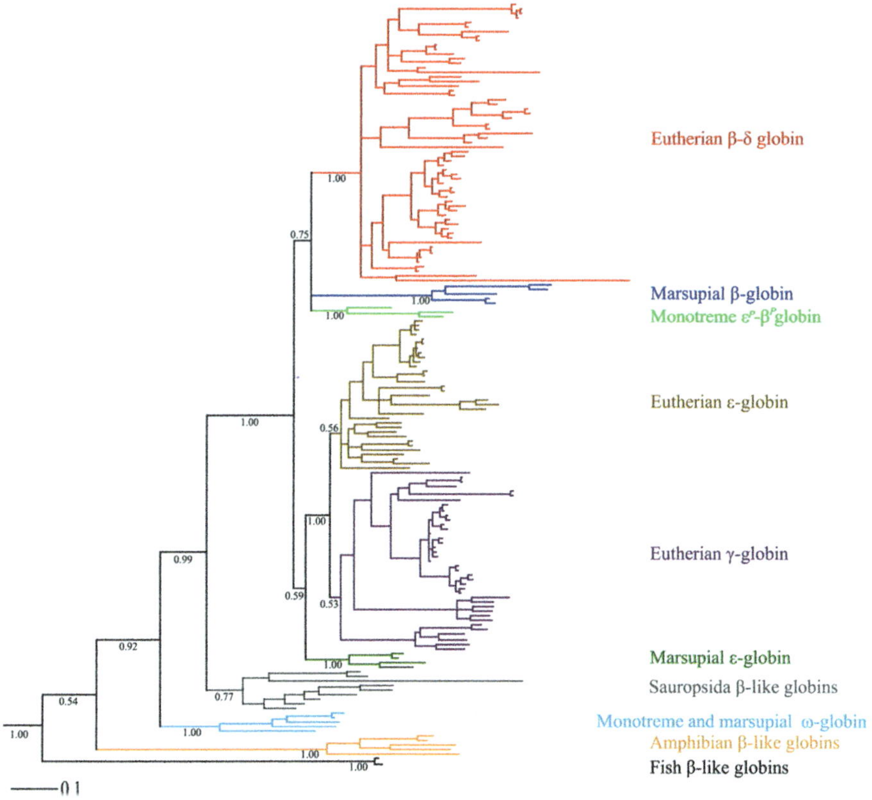

Fig. 10.6 Phylogenetic tree showing relationships among the β-like globin genes of vertebrates. (Source: Opazo et al. 2008)

Fig. 10.7 Phylogenetic tree showing relationships among land plants, archae, and bacterial V-PPases. (Source: Suneetha et al. 2016)

antiporter and helps in Na$^+$ compartmentation. Suneetha et al. (2016) carried out phylogenetic studies on land plants, archaea, and bacterial V-PPases (Fig. 10.7). V-PPases are highly conserved among land plants and less among archaeon, protozoan, and bacteria (Suneetha et al. 2016). Phylogeny with respect to other land plants revealed that V-PPases of *A. thaliana* (AtVPP), *H. vulgare* (HvVPP), *B. vulgaris* (BvVPP), *N. tabacum* (NtVPP), and *O. sativa* (OsVPP) are highly conserved.

10.1.5   Classifying Proteins or Genes into Families

Classification of genes into gene families is important for understanding function and evolution of gene. There are three methods to infer gene families: (1) using phylogenetic trees for classification, (2) using similarities with known sequence signatures like motifs or domains, and (3) pairwise comparisons involving the use of clustering techniques (Frech and Chen 2010).

Phylogenetic tree was used for effective classification of ABC transporter gene families. Multiple sequence alignment of both known and putative new ABC transporter family C genes using ClustalW with default parameters was performed. The phylogenetic tree was produced by the minimum evolution method and 1000 bootstrap iteration. In phylogenetic analysis, the three new genes grouped nicely within known ABC transporters of family C (Fig. 10.8). Thus, phylogenetic analysis can be used to classify new genes into ABC transporter family C (Frech and Chen 2010).

10.1.6 Understanding the History of Life

Understanding the systematics of living organisms in the world is a challenging task. Literature reports several studies carried out to understand the kingdom-level phylogeny. Carl Woese established a molecular sequence-based phylogenetic tree by comparing ribosomal RNA (rRNA) sequences that could relate all organisms and reconstruct the history of life (Woese 1987; Woese and Fox 1997). Woese articulated and recognized three primary lines of evolutionary descent, termed "urkingdoms" or "domains": .....“Eucarya (eukaryotes), Bacteria (eubacteria), and Archaea (archaebacteria)”..... (Woese et al. 1990). Pace (1997) used molecular phylogeny to compile the robust map of life domains: Archaea, Bacteria, and Eucarya (Fig. 10.9). The universal phylogenetic tree based on 64 SSU rRNA sequences was aligned, and a tree was produced using FASTDNAML. Baldauf et al. (2000) used concatenated amino acid sequences of four protein-encoding genes to produce a phylogenetic tree for 14 higher-order eukaryote taxa (Fig. 10.10). Thus, phylogeny was used to understand the kingdom-level relations.

10.1.7 Estimating the Time of Divergence Using Molecular Clock

Molecular dating techniques were used to estimate the time of species divergences. Literature reports several research studies used to determine the time of species divergences. Molecular dating requires standard sequence datasets; statistical distributions to model; and prior divergence times to find out the time of divergence during the course of evolution. Hasegawa et al. (1985) developed a method for estimating divergence dates of humans from species by a molecular clock approach. The molecular clock of mitochondrial DNA (mt DNA) was calibrated ~65 million years ago and a generalized least squares method was applied. The divergence dates were 92.3 ± 11.7, 13.3 ± 1.5, 10.9 ± 1.2, 3.7 ± 0.6, and 2.7 ± 0.6 million years ago for mouse, gibbon, orangutan, gorilla, and chimpanzee, respectively (Figs. 10.11 and 10.12). Thus, phylogeny can be used to estimate time of divergence for species of interest.

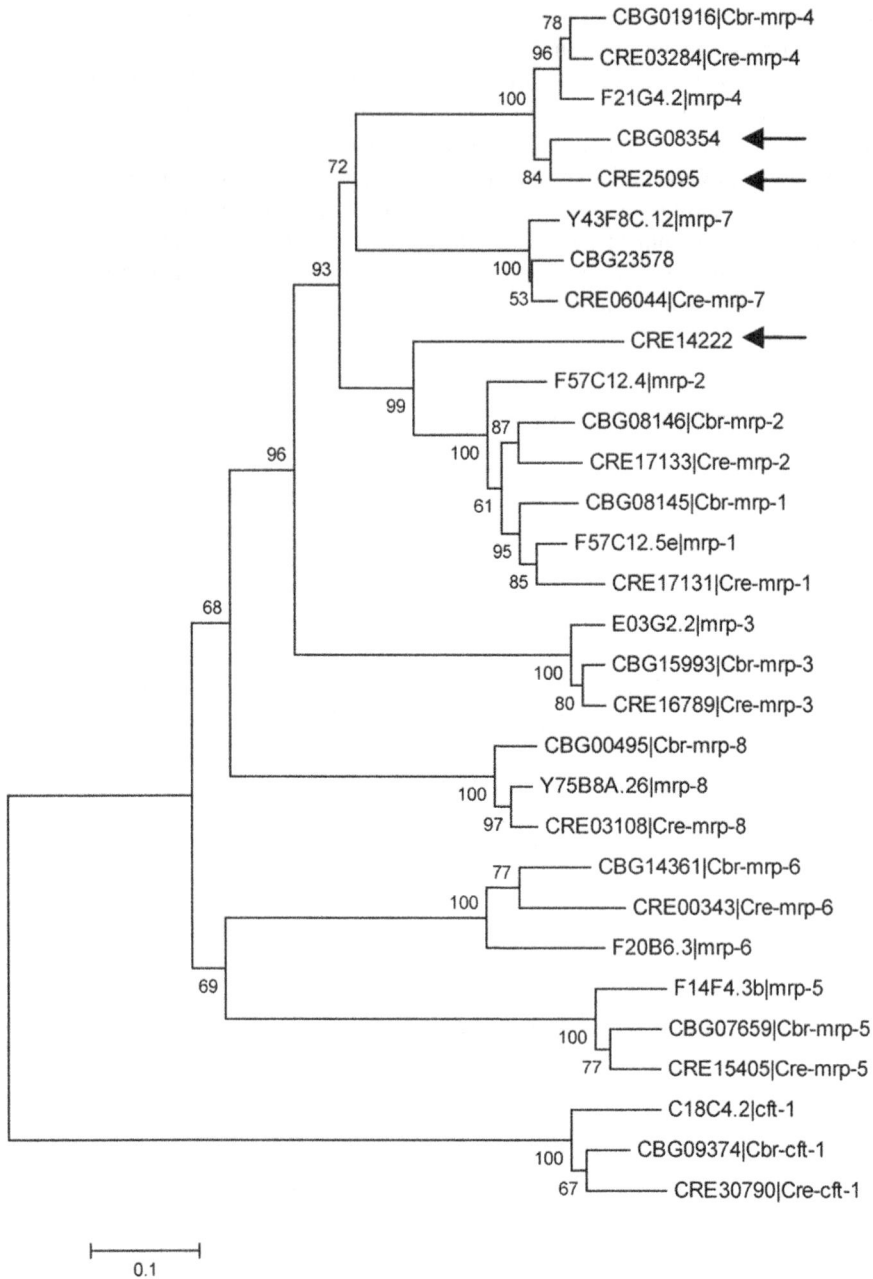

Fig. 10.8 The phylogenetic tree shows the evolutionary relationship of the three new ABC transporter genes CBG08354, CRE25095, and CRE14222 (indicated by arrows) with known *C. elegans*, *C. briggsae*, and *C. remanei* ABC transporters of family C. (Source: Frech and Chen 2010)

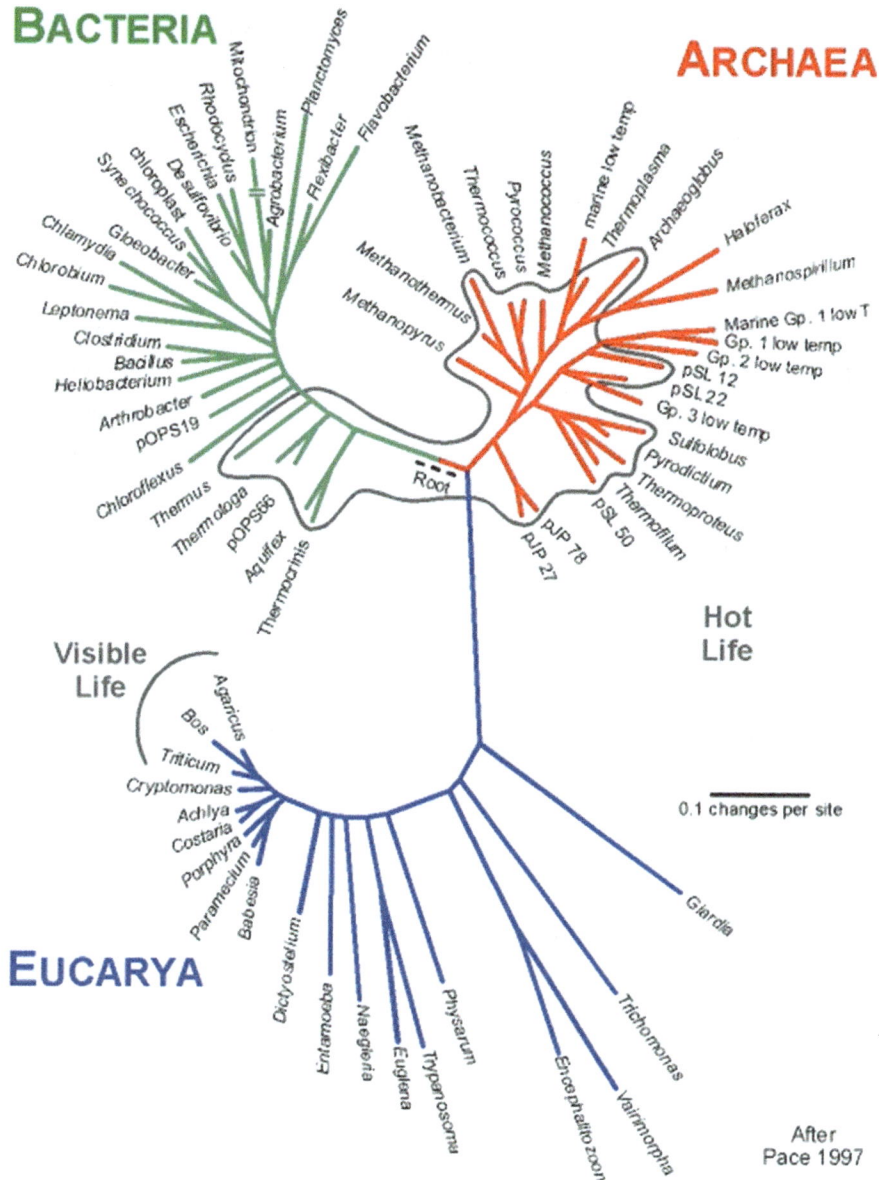

Fig. 10.9 The phylogenetic tree shows the robust map of life domains: Archaea, Bacteria, and Eucarya. (Source: Pace 1997)

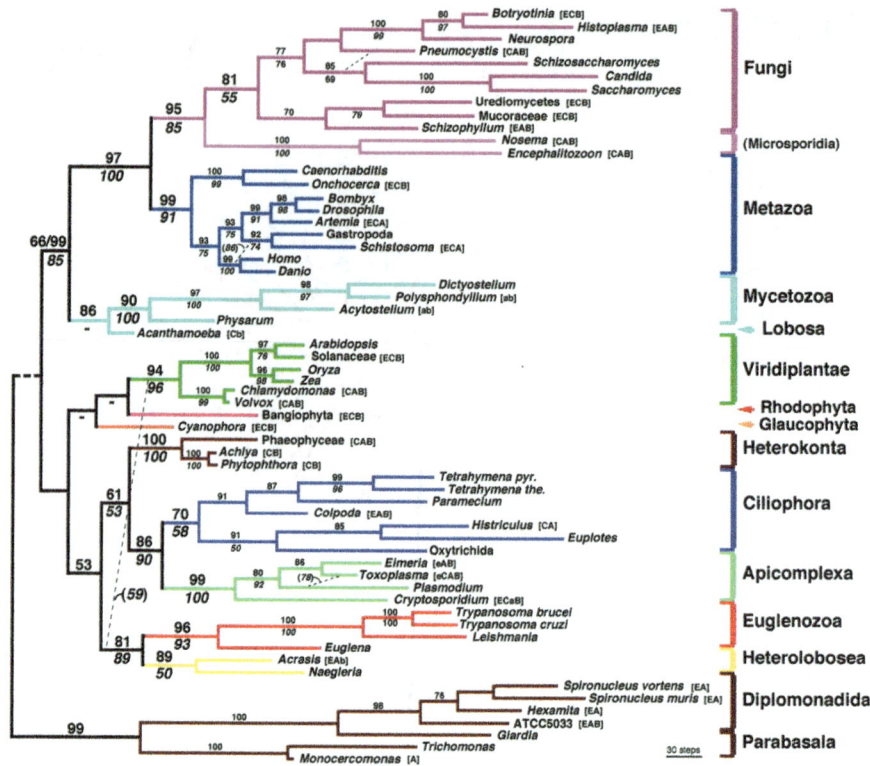

Fig. 10.10 The phylogenetic tree shows the 14 higher-order eukaryote taxa. (Source: Baldauf et al. 2000)

10.1.8 Evolution of Pathogen

Viruses are with high mutation rate and adapt quickly to environmental changes leading to the high genetic diversity. On the other hand, this fast evolution leaves behind significant marks in the genome of virus that can be connected with transmission dynamics and epidemiology. Evolutionary theory and sequence analysis played a role in understanding epidemiology of virus by figuring out the origin of time and geographical site of a virus. Analysis was able to provide information on transmission linkages or chains for a population.

Huet et al. (1990) inferred the origin and classified HIV into types, groups, and subtypes (Fig. 10.13). Epidemiological, physiological, and clinical evidences favored cross-species transmission of HIV from chimpanzee to humans (Castro-Nallar et al. 2012). Further, phylogenetic evidence corroborates this fact that HIV-1 and HIV-2 are due to several cross-species transmission events (Huet et al. 1990; Gao et al. 1992, 1999; Hahn et al. 2000; Plantier et al. 2009; Van Heuverswyn and Peeters 2007) (Fig. 10.14).

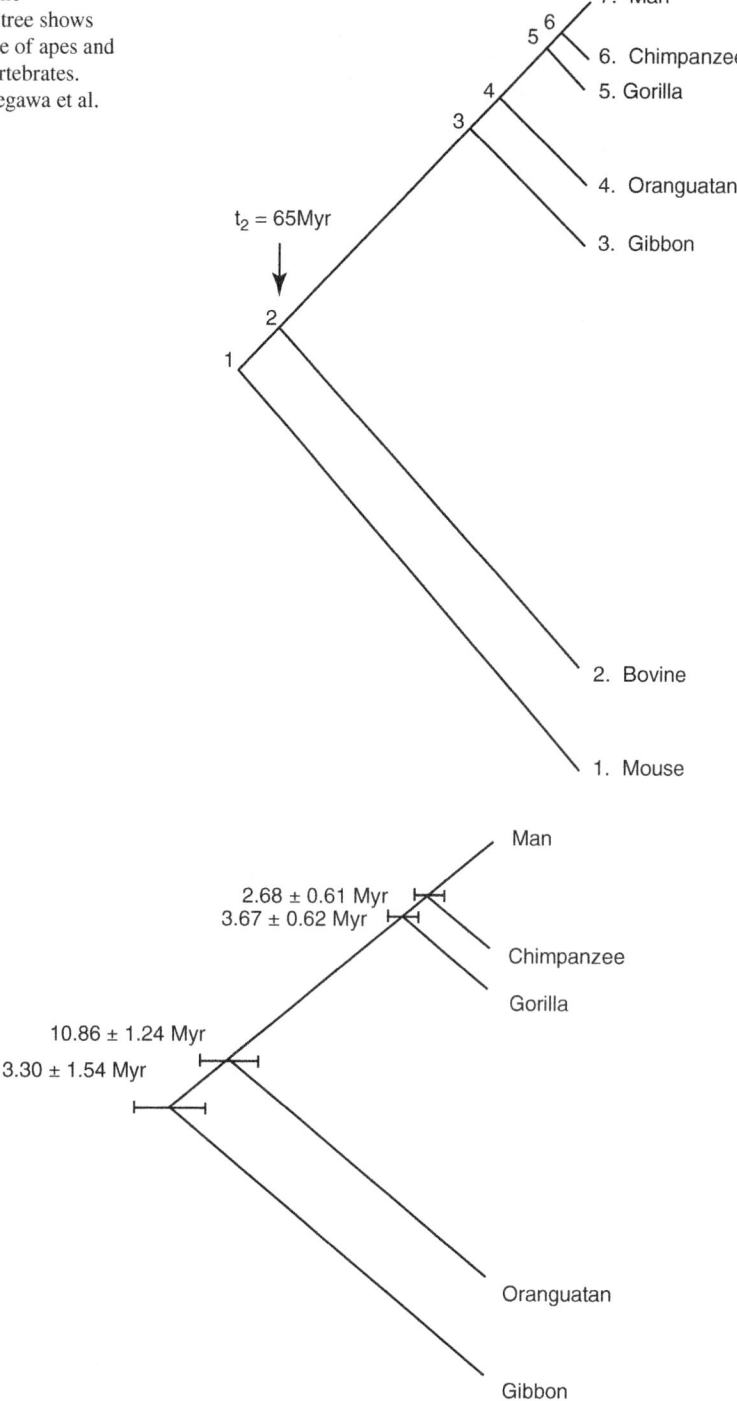

Fig. 10.11 The phylogenetic tree shows the divergence of apes and humans in vertebrates. (Source: Hasegawa et al. 1985)

$t_2 = 65\text{Myr}$

7. Man
6. Chimpanzee
5. Gorilla
4. Oranguatan
3. Gibbon
2. Bovine
1. Mouse

2.68 ± 0.61 Myr
3.67 ± 0.62 Myr

10.86 ± 1.24 Myr
13.30 ± 1.54 Myr

Man
Chimpanzee
Gorilla
Oranguatan
Gibbon

Fig. 10.12 The phylogenetic tree shows the divergence of humans from apes. (Source: Hasegawa et al. 1985)

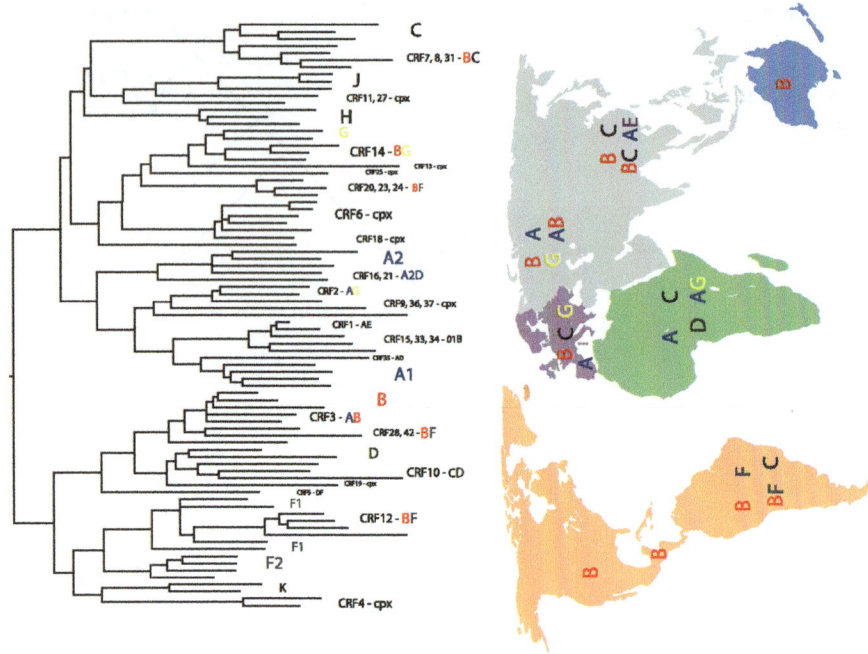

Fig. 10.13 Phylogenetic tree representation of HIV-1 and its subtypes. (Source: Castro-Nallar et al. 2012)

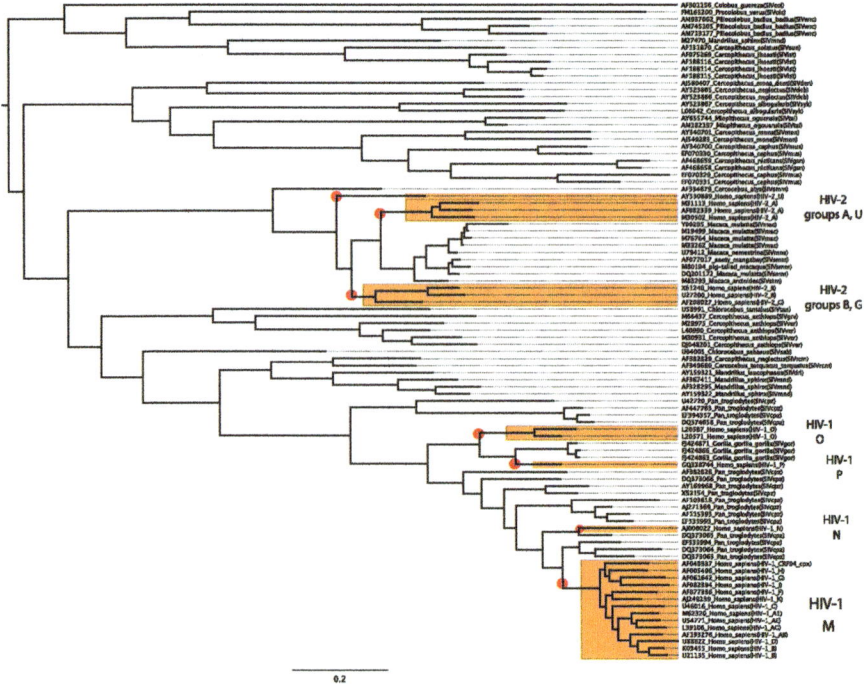

Fig. 10.14 Phylogenetic tree showing HIV cross-species transmission. (Source: Castro-Nallar et al. 2012)

Intensive studies were carried out on the evolution and divergence of HIV-1 and HIV-2 using phylogeny. The divergence time of HIV-1, HIV-2 (subtype A), and HIV-2 (subtype B) dated to the 1920s (Worobey et al. 2008), 1940 ± 16 (Lemey et al. 2003), and 1945 ± 14 (Lemey et al. 2003), respectively. Introduction of clade B of HIV-1 into North America dated to 1968 (1966–1970) (Gilbert et al. 2007; Pérez-Losada et al. 2010).

The emerging field of phylodynamics—"the melding of immunodynamics, epidemiology, and evolutionary biology …"—was used to understand the transmission dynamics, population dynamics, and within-host dynamics of virus or bacteria (Grenfell et al. 2004). Transmission dynamics helps in understanding diversity of an organism in transmission network constructed during a transmission event for potential therapy development. Population dynamics increases our understanding on patterns of diversity among populations throughout the length and breadth of infection, within host and transmission events. Within-host dynamics provide information on evolution of virus in the host which is associated with disease progression. There are two aspects within host dynamics which are observed in case of HIV. The first one is that evolution of HIV is different in specific tissues. It was revealed that HIV evolves at different rates in different compartments of the brain, which cannot be attributed to selective pressure, but can be related to viral expansion due to immune failure (Salemi et al. 2005). The second aspect is that HIV genetic diversity (variation) in the host leads to evolution of quasispecies (Holmes 2009). So, phylodynamics can be useful in relating epidemiological and evolutionary information which can be used for monitoring surveillance programs of a virus especially in case of HIV. Thus, phylogenetics can be used to identify evolution of virus in terms of origin, time of divergence, pathogen evolution, and understand phylodynamics.

10.2 Construction of Phylogenetic Trees

Data and tree construction methods used for construction of phylogenetic tree effect topology of the tree; therefore, it is worth to discuss on data and tree construction methods.

10.2.1 Data

Data generated via fingerprinting techniques such as rapid amplification polymorphic DNA (RAPD), restriction fragment length polymorphism (RFLP), AFLP, SSCP, and sequence data (nucleotide and protein sequence data) are used for phylogeny. Data from fingerprinting techniques such RAPD, AFLP, and SSCP is converted to binary data (0/1). The "0s" represent the absence of band in the DNA fingerprinting techniques, whereas "1s" represent the presence of band in the DNA fingerprinting techniques. DNA or protein sequence data is generated by Sanger's method.

This binary or sequence data is either converted to distance or used directly in the form of character used to construct a phylogenetic tree. The fingerprinting data or the sequence data (DNA or protein) was known to influence the tree topology of the phylogenetic tree (Neelapu 2007; Devi et al. 2006, 2007; Padmavathi et al. 2003).

10.2.2   Tree-Constructing Methods

Broadly, there are two fundamental methods for constructing phylogenetic trees: distance or discrete character methods. Distance methods first convert data or aligned sequences into pairwise distance matrix. A correction is needed for these raw distances. These corrections are based on the assumptions of various substitution models proposed for both nucleic acid and protein sequence methods. A phylogenetic tree building method is then used to construct an evolutionary tree. Some of the tree-building methods are unweighted pair group method with arithmetic means (UPGMA), minimum evolution, neighbor joining, and Fitch-Margoliash.

UPGMA (Sokal and Michener 1958; Nei 1975) clusters data based on similarity and assumes that changes are accumulated at a constant rate among the lineages. In neighbor-joining method (Saitou and Nei 1987), a star tree in which terminal taxa are equidistant, is first established; then, two taxa are temporarily taken from the star to a new node, and the total distance in the new tree is recalculated; and the taxa are returned to the star and another pair of taxa is taken to repeat the operation. This process is continued until all the taxa are jointed in a completely resolved tree with the lowest total distance. In minimum evolution method (Takahashi and Nei 2000), the initial tree is created by clustering taxa using neighbor-joining method. Then, every possible tree is examined and one tree with minimum branch length is selected, thereby minimizing the total distance in a tree.

Discrete methods directly consider the state of each nucleotide or amino acid site in each sequence under comparison. The two discrete character methods are maximum likelihood and maximum parsimony. Maximum likelihood method (Cavalli-Sforza and Edwards 1967; Felsenstein 1973; Felsenstein 1981; Swofford et al. 1996) uses data to determine the probability of substitution, relative frequencies, and the different probabilities of transitions and transversions. It then selects the tree that maximizes the probability of good fit of the data. Maximum likelihood method presents an additional opportunity to evaluate trees with variations in mutation rates in different lineages; and also to use explicit evolutionary models such as the jukes-cantor and Kimura models.

Parsimony is another discrete character method that creates evolutionary trees based on a systematic search among possible trees for the fewest plausible mutational steps from a common ancestor necessary to account for two diverged lineages, and those trees that require the fewest changes are said to be most parsimonious (i.e., optimal) trees. The sum of the minimum possible substitutions over all sites is known as the tree length for that topology. The topology with the minimum tree length is known as the maximum parsimony tree. Three different types of searches the max-mini branch-and-bound search, min-mini heuristic search, and close-neighbor-interchange heuristic search are performed to generate maximum parsimony tree. The maximum parsimony method (Fitch 1971) produces many equally parsimonious trees. A majority-rule consensus method is used to produce a composite tree that is a consensus among all such trees.

10.2.3 Phylogeny Program Packages

All these clustering methods are available in various phylogenetic packages such as PHYLIP (Felsenstein 1989), PAUP (Swofford 1991), MEGA (Kumar et al. 2004), TreePuzzle (Schmidt et al. 2002), etc. (Table 10.1). The computational limits that were faced in running maximum parsimony and maximum likelihood method with increase in number of species and increase in length of the sequence in most packages are overcome in MEGA. Moreover, best tree editing options such as Tree Explorer program are available in MEGA, which makes phylogenetic inference from sequence data much easier.

10.3 Methods to Assess the Confidence of Phylogenetic Tree

The tree generated based on the input data and tree construction method is known as inferred tree. This inferred tree need not be the true tree for the given phylogenetic data. So, there is a requirement to test the reliability of the phylogenetic tree or portion of the tree. In methods like minimum evolution, maximum parsimony, and maximum likelihood, increase in tree number is observed as the sample size increases (Table 10.2). In these conditions, whether the tree is significant/better than another tree is to be confirmed. The reliability of the phylogenetic tree or portion of the tree is tested by sampling methods, whereas the significant difference of a tree over the other is confirmed by statistical tests.

10.3.1 Sampling Methods

The reliability of the phylogenetic tree or portion of the tree is tested by sampling methods such as bootstrapping, jackknifing, and Bayesian simulation.

10.3.1.1 Bootstrapping

Bootstrapping is random sampling with replacement of data (distances or sequence: nucleotide or protein) which addresses if any sampling errors occurred for the required analysis. In molecular phylogeny, bootstrapping repeatedly samples the

Table 10.1 Phylogeny packages and resources available for phylogenetic analysis

S. No	Phylogeny packages	Description	Resource available at	References
1	**PHYL**ogeny Inference **P**ackage (PHYLIP)	PHYLIP is a free package available for inferring phylogenies. Molecular sequences; gene frequencies; restriction sites and fragments; distance matrices; and discrete characters are used as input. Parsimony, distance, and likelihood methods are used to construct phylogenetic tree	http://evolution.genetics. washington.edu/phylip/ general.html	Felsenstein (1989)
2	Phylogenetic Analysis Using Parsimony (PAUP)	Phylogenetic relationships are performed for molecular, morphological, and behavioral data. The data for PAUP is NEXUS format. PAUP is used in the field of evolutionary biology, conservation biology, ecology, and forensic studies	http://paup.sc.fsu.edu/ about.html	Swofford (1991)
3	MacClade's	MacClade is a phylogenetic analysis suite which is used for studying evolution of characters-based methods. The features include entering and editing data and phylogenies	http://macclade.org/index. html	Maddison and Maddison (1992)
4	FastDNAml	FastDNAml constructs phylogenetic tree for DNA sequences based on maximum likelihood method with improved performance	http://iubio.bio.indiana.edu/ soft/molbio/evolve/ fastdnaml/fastDNAml.html	Olsen et al. (1994)
5	**MOL**ecular **PHY**logenetics (Molphy)	MolPhy infers phylogeny of nucleic acid and protein using maximum likelihood	http://www.softpedia.com/ get/Science-CAD/Molphy. shtml	Adachi and Hasegawa (1996)
6	GeneTree	GeneTree uses parsimony method to produce gene trees and species phylogenies	http://taxonomy.zoology. gla.ac.uk/rod/genetree/ genetree.html	Page (1998)
7	Tree analysis using New Technology (TNT)	TNT is a parsimonious method used for searching parsimony tree	http://www.zmuc.dk/ public/phylogeny/TNT	Goloboff (1999)
8	TCS	TCS estimates gene genealogies within a population in a minimum spanning tree (rooted network tree) based on parsimony method. The accepted input data is sequence or absolute distances in NEXUS or PHYLIP format	http://darwin.uvigo.es/ software/tcs.html	Clement et al. (2000)
9	GEODIS	GEODIS analyzes haplotypes in a genealogy to distinguish between historical divergence of populations and geographical separation	http://darwin.uvigo.es/ software/geodis.html	Posada et al. (2000)

	Name	Description	URL	Reference
9	Phylogenetic Analysis by Maximum Likelihood (PAML)	PAML analyzes DNA or protein sequences using maximum likelihood	http://abacus.gene.ucl.ac.uk/software/paml.html	Yang (2000)
10	PAL (Phylogenetic Analysis Library)	PAL facilitates construction of phylogenetic trees based on maximum likelihood, neighbor-joining, and least squares methods. PAL assesses the confidence of tree by bootstrapping, Kishino-Hasegawa-Templeton, and Shimodaira-Hasegawa tests. The divergence time in a tree can also be estimated	http://www.cebl.auckland.ac.nz/pal-project/	Drummond and Strimmer (2001)
11	Mr Bayes	Bayesian phylogenetic inference is based on Markov chain Monte Carlo (MCMC) methods	http://mrbayes.sourceforge.net/	Huelsenbeck and Ronquist (2001)
12	Data Analysis in Molecular Biology and Evolution (DAMBE)	DAMBE infers phylogeny from DNA and protein sequence by parsimony, distance, and likelihood methods. It can assess the confidence of tree either by bootstrapping or jackknifing. It can read and convert a number of file formats	http://dambe.bio.uottawa.ca/dambe.asp	Xia and Xie (2001)
13	TreePuzzle	TreePuzzle constructs phylogenetic tree from sequence data by maximum likelihood method. The confidence of the tree can be assessed by Kishino-Hasegawa test and Shimodaira-Hasegawa test	http://www.tree-puzzle.de/	Schmidt et al. (2002)
14	Molecular Evolutionary Genetic Analysis (MEGA)	Sophisticated and user-friendly software suite for sequence analysis, tree construction, and data visualization. Sequence Analyses suite includes phylogeny inference, dating and clocks, recognizing ancestral states, sequence alignment for DNA, and protein sequence data. The statistical methods maximum likelihood, distance methods, ordinary least squares, maximum parsimony, and composite likelihood are used for construction of tree. The powerful data visual tools are Alignment/Trace Editor, Tree Explorer, Data Explorers, Legend Generator, Gene Duplication Wizard, and Timetree Wizard	https://www.megasoftware.net/	Kumar et al. (2004)
15	IQPNNI	An efficient tree reconstruction method which is based on important quartet puzzling (IQP) and nearest neighbor interchange (NNI) operation	http://www.cibiv.at/software/iqpnni/	Vinh and von Haeseler (2004)
16	BayesPhylogenies	BayesPhylogenies is an application based on Bayesian Markov Chain Monte Carlo (MCMC) or Metropolis-coupled Markov chain Monte Carlo (MCMCMC) method which models the evolution	http://www.evolution.reading.ac.uk/BayesPhy.html	Pagel and Meade (2004)

(continued)

Table 10.1 (continued)

S. No	Phylogeny packages	Description	Resource available at	References
17	BAli-Phy	BAli-Phy is a phylogenetic application that generates multiple sequence alignments and evolutionary trees based on likelihood method from DNA and amino acid sequences	http://www.bali-phy.org/	Suchard and Redelings (2006)
18	BUCKy	BUCKy combines molecular data from multiple loci of sampled individuals to analyze the assumption of Bayesian concordance or discordance among gene trees	https://www.stat.wisc.edu/~ane/bucky/	Ané et al. (2007)
19	Bosque	Phylogenetics application for sequence management, alignment, and construction of phylogenetic tree	http://bosque.udec.cl/	Ramírez and Ulloa (2008)
20	FastTree	FastTree infers phylogenetic trees from nucleotide or protein sequences using maximum-likelihood method 1000 times faster than regular maximum-likelihood applications with reasonable time and memory	http://www.microbesonline.org/fasttree/	Price et al. (2009)
21	MetaPIGA	MetaPIGA infers phylogeny under maximum likelihood for binary and character data (nucleotide and protein)	http://www.metapiga.org/	Raphaël and Milinkovitch (2010)
22	Mesquite	Mesquite is modular software for evolutionary biology to help biologists in population genetics, phylogenetic analysis, and geological timescale analysis	http://mesquiteproject.org/	Maddison and Maddison (2011)
23	Bayesian Evolutionary Analysis Sampling Trees (BEAST)	Bayesian analysis based on MCMC method infers phylogenetic trees	http://beast.community/	Drummond et al. (2012)
24	Armadillo	Armadillo is an open-source computer applications suite for bioinformatics analysis. The following bioinformatics tasks are performed using Armadillo—"automatic BLAST queries, multiple sequence alignment, evolutionary model inference, construction of phylogenetic trees (distance, maximum likelihood, maximum parsimony, and Bayesian methods), visualization of phylogenetic trees and networks, phylogenetic tree manipulation, and horizontal gene transfer detection"	http://www.bioinfo.uqam.ca/armadillo/	Lord et al. (2012)
25	IQ-TREE	A fast and efficient maximum likelihood method to infer phylogenetic tree from the data	http://www.iqtree.org/	Nguyen et al. (2015)
26	TOPALi	TOPALi estimates phylogenetic tree based on Bayesian analysis (BA) and maximum likelihood (ML)	http://www.topali.org/	McGuire and Wright (2000)

Table 10.2 The number of rooted trees and unrooted trees for n sequences

Number of taxa	Number of unrooted trees Formula $Nu = \dfrac{(2n-5)!}{2^{n-3}(n-3)!}$	Number of rooted trees $Nr = \dfrac{(2n-3)!}{2^{n-2}(n-2)!}$
3	1	3
4	3	15
5	15	105
6	105	945
7	945	10,395
8	10,395	135,135
9	135,135	2,027,025
10	2,027,025	34,459,425

data to construct the phylogenetic tree and gives us the chance to assess the strength of the original tree. If the data resampling generates different trees when compared with the original tree, then the tree topology is based on the data with weak phylogenetic signals. If the data resampling generates tree similar to the original tree, then the tree topology is based on the data with enough phylogenetic signals. Thus, bootstrapping (resampling data) provides insights on the confidence of the tree topology.

Two types of bootstrapping are used in phylogenetic analysis: parametric or nonparametric bootstrapping. If the data is disturbed by random sampling generating new dataset, then it is nonparametric bootstrapping. If the data is disturbed by particular order to generate new dataset, then it is parametric bootstrapping. The other types of bootstrapping are case resampling, Bayesian bootstrap, smooth bootstrap, resampling residuals, Gaussian process regression bootstrap, wild bootstrap, and block bootstrap (time series: simple block bootstrap, time series: moving block bootstrap, cluster data: block bootstrap).

If bootstrapping is repeated 100–1000 times or even more to reconstruct phylogenetic trees, then certain parts of the tree have different topology when compared with the original inferred tree. All these bootstrapped trees are summed up into a consensus tree based on a majority rule. The most supported branching patterns shown at each node are labeled with bootstrap values. Thus, bootstrap offers a measure for estimating the confidence levels of the tree topology.

10.3.1.2 Jackknifing

Jackknifing is another resampling technique where half of the dataset is randomly deleted, generating datasets half-original. Initially, a phylogenetic tree is constructed with the original dataset, then with each new dataset generated by jackknifing, a phylogenetic tree is constructed using the same method as the original. Sampling generates different trees when phylogenetic signals are weak, whereas sampling generates similar tree when phylogenetic signals are strong. Thus, jackknifing (resampling data) can also be used to assess the confidence of the tree topology.

10.3.1.3 Bayesian Method

Bayesian method based on MCMC approach resamples data thousands or millions of steps or iterations. The sample datasets are used to reconstruct phylogenetic trees similar to original inferred tree. The posterior probabilities designated at each intersection of a best Bayesian tree measure the confidence levels of the tree topology.

10.3.2 Statistical Methods

The significant difference of a tree over the other is confirmed by statistical tests such as Kishino-Hasegawa Test and Shimodaira-Hasegawa Test.

10.3.2.1 Kishino-Hasegawa Test

Kishino-Hasegawa (KH) test compares two tree topologies to differentiate one tree over the other (Kishino and Hasegawa 1989). Though KH test can be used for differentiating trees generated through methods such as distance, parsimony, and likelihood, Kishino-Hasegawa developed this test specifically for parsimonious trees. The KH test (statistical method) is paired Student t-test based on null hypothesis that the "two competing tree topologies are not significantly different…." The standard deviation of the difference between branch lengths at each informative site between two trees is estimated. Then the derived t-value is compared with the t-distribution values either to accept or reject the null hypothesis at certain significant levels (with probability e.g., $P < 0.05$).

$$t = \frac{Da - Dt}{SD / \sqrt{n}}$$

df = $(n - 1)$ where t is the test statistical value, Da is the average site-to-site difference between the two trees, Dt is the total difference of branch lengths of the two trees, SD is the standard deviation, n is the number of informative sites, and df is the degree of freedom.

10.3.2.2 Shimodaira-Hasegawa Test

Shimodaira-Hasegawa (SH) developed a statistical test for ML trees based on likelihood ratio using the $\chi2$ test to estimate the goodness of fit of two competing trees (Shimodaira and Hasegawa 1999). The log likelihood scores lnLA and lnLB for tree A and tree B are obtained first, for the two competing trees. Then the log ratio of the two scores is obtained by $d = 2(\ln LA - \ln LB) = 2 \ln (LA/LB)$ and used to test against the $\chi 2$ distribution from a table. The resulting probability value (P-value) determines whether the difference between the two trees is significant or nonsignificant.

10.4 Conclusion

Molecular phylogeny establishes the relationships among the set of objects in the study. Binary data ("0"/"1") from RAPD, RFLP, AFLP, SSCP, and sequence data (DNA or protein) from the set of objects are used to construct phylogenetic tree. The different tree construction methods are UPGMA, NJ, ME, FM, MP, and ML. Molecular phylogeny has a wide range of applications and if the interpretation of the evolutionary patterns is not appropriate, then the inference of the study may be misleading. The interpretation of the tree is always dependent on assessing the confidence of the phylogenetic tree. Sampling methods (bootstrapping, jackknifing, and Bayesian simulation) and statistical methods (KH test and SH test) can be used to assess the confidence of the phylogenetic tree. Thus, if the confidence of the phylogenetic tree generated is good, then the interpretation or inference of the study will not be misleading.

Acknowledgment The authors are grateful to Gandhi Institute of Technology and Management (GITAM) Deemed-to-be-University, for providing necessary facilities to carry out the research work and for extending constant support in writing this review.

References

Adachi J, Hasegawa M (1996) Molphy, version 2.3. Programs for molecular phylogenetics based on maximum likelihood. In: Ishiguro M, Kitagawa G, Ogata Y, Takagi H, Tamura Y, Tsuchiya T (eds) Computer science monographs. Institute of Statistical Mathematics, Tokyo

Ané C, Larget B, Baum DA, Smith SD, Rokas A (2007) Bayesian estimation of concordance among gene trees. Mol Biol Evol 24(2):412–426

Baldauf SL, Roger AJ, Wenk-Siefert I, Doolittle WF (2000) A kingdom-level phylogeny of eukaryotes based on combined protein data. Science 290(5493):972–977

Castro-Nallar E, Perez-Losada M, Burton GF, Crandall KA (2012) The evolution of HIV: inferences using phylogenetics. Mol Phylogenet Evol 62:777–792

Cavalli-Sforza LL, Edwards AWF (1967) Phylogenetic analysis: models and estimation procedures. Evolution 21:550–570

Clement M, Posada D, Crandall K (2000) TCS: a computer program to estimate gene genealogies. Mol Ecol 9:1657–1660

Devi KU, Reineke A, Reddy NNR, Rao CUM, Padmavathi J (2006) Genetic diversity, reproductive biology, and speciation in the entomopathogenic fungus *Beauveria bassiana* (Balsamo) Vuillemin. Genome 49(5):495–504

Devi UK, Reineke A, Rao UCM, Reddy NRN, Khan APA (2007) AFLP and single strand conformation polymorphism studies of recombination in the entomopathogenic fungus *Nomuraea rileyi*. Mycol Res 111(6):716–725

Drummond A, Strimmer K (2001) PAL: an object-oriented programming library for molecular evolution and phylogenetics. Bioinformatics 17:662–663

Drummond AJ, Suchard MA, Xie D, Rambaut A (2012) Bayesian phylogenetics with BEAUti and the BEAST 1.7. Mol Biol Evol 29:1969–1973

Felsenstein J (1973) Maximum likelihood and minimum-steps methods for estimating evolutionary trees from data on discrete characters. Syst Zool 22:240–249

Felsenstein J (1981) Evolutionary trees from gene-frequencies and quantitative characters – finding maximum-likelihood estimates. Evolution 35:1229–1242

Felsenstein J (1989) PHYLIP – phylogeny inference package (version 3.2). Cladistics 5:164–166

Fitch WM (1971) Towards defining the course of evolution: minimum change for a specific tree topology. Syst Zool 20:406–416

Frech C, Chen N (2010) Genome-wide comparative gene family classification. PLoS One 5(10):e13409. https://doi.org/10.1371/journal.pone.0013409

Gao F, Yue L, White AT, Pappas PG, Barchue J, Hanson AP, Greene BM, Sharp PM, Shaw GM, Hahn BH (1992) Human infection by genetically diverse SIVSM-related HIV-2 in West Africa. Nature 358:495–499

Gao F, Bailes E, Robertson DL, Chen Y, Rodenburg CM, Michael SF, Cummins LB, Arthur LO, Peeters M, Shaw GM, Sharp PM, Hahn BH (1999) Origin of HIV-1 in the chimpanzee *Pan troglodytes troglodytes*. Nature 397:436–441

Gilbert MTP, Rambaut A, Wlasiuk G, Spira TJ, Pitchenik AE, Worobey M (2007) The emergence of HIV/AIDS in the Americas and beyond. Proc Natl Acad Sci U S A 104:18566–18570

Goloboff PA (1999) Analyzing large data sets in reasonable times: solutions for composite optima. Cladistics 15:415–428

Grenfell B, Pybus O, Gog J, Wood J, Daly J (2004) Unifying the epidemiological and evolutionary dynamics of pathogens. Science 303:327–332

Hahn BH, Shaw GM, De Cock KM, Sharp PM (2000) AIDS as a zoonosis: scientific and public health implications. Science 287:607–614

Hardison RC (2012) Evolution of hemoglobin and its genes. Cold Spring Harb Perspect Med 2(12):a011627. https://doi.org/10.1101/cshperspect.a011627

Hasegawa M, Kishino H, Yano T (1985) Dating of the human-ape splitting by a molecular clock of mitochondrial DNA. J Mol Evol 22(2):160–174

Holmes EC (2009) The evolution and emergence of RNA viruses. Oxford University Press, New York

Huelsenbeck JP, Ronquist F (2001) MrBayes: Bayesian inference of phylogeny. Bioinformatics 17:754–755

Huet T, Cheynier R, Meyerhans A, Roelants G, Wain-Hobson S (1990) Genetic organization of a chimpanzee lentivirus related to HIV-1. Nature 345:356–359

Kishino H, Hasegawa M (1989) Evaluation of the maximum likelihood estimate of the evolutionary tree topologies from DNA sequence data, and the branching order in Hominoidea. J Mol Evol 29:170–179

Kumar S, Tamura K, Nei M (2004) MEGA3: an integrated software for molecular evolutionary genetics analysis and sequence alignment. Brief Bioinform 5:150–163

Lemey P, Pybus OG, Wang B, Saksena NK, Salemi M, Vandamme A-M (2003) Tracing the origin and history of the HIV-2 epidemic. Proc Natl Acad Sci U S A 100:6588–6592

Lord E, Leclercq M, Boc A, Diallo AB, Makarenkov V (2012) Armadillo 1.1: an original workflow platform for designing and conducting phylogenetic analysis and simulations. PLoS One 7(1):e29903. https://doi.org/10.1371/journal.pone.002990

Maddison WP, Maddison DR (1992) MacClade. Sinauer Associates, Sunderland

Maddison WP, Maddison DR (2011) Mesquite: a modular system for evolutionary analysis. Version 2.75. http://mesquiteproject.org

Maeshima M (2000) Vacuolar H+-pyrophosphatase. Biochim Biophys Acta 1465:37–51

Margos G, Vollmer SA, Ogden NH, Fish D (2011) Population genetics, taxonomy, phylogeny and evolution of *Borrelia burgdorferi* sensu lato. Infect Genet Evol 11(7):1545–1563

McGuire G, Wright F (2000) TOPAL 2.0: improved detection of mosaic sequences within multiple alignments. Bioinformatics 16(2):130–134

Neelapu NRR (2007) Investigation on existence and mechanism of recombination and molecular phylogeny of mitosporic entomopathogenic fungi *Beauveria bassiana* (Balsamo) Vuillemin and *Nomuraea rileyi* (Farlow) Samson. Doctoral dissertation, Andhra University, Visakhapatnam, India

Neelapu NRR, Reineke A, Chanchala UMR, Koduru UD (2009) Molecular phylogeny of asexual entomopathogenic fungi with special reference to *Beauveria bassiana* and *Nomuraea rileyi*. Rev Iberoam Micol 26(2):129–145

Nei M (1975) Molecular population genetics and evolution. North-Holland, Amsterdam

Nguyen L-T, Schmidt HA, von Haeseler A, Minh BQ (2015) IQ-TREE: a fast and effective stochastic algorithm for estimating maximum-likelihood phylogenies. Mol Biol Evol 32(1):268–274

Olsen GJ, Matsuda H, Hagstrom R, Overbeek R (1994) FastDNAml: a tool for construction of phylogenetic trees of DNA sequences using maximum likelihood. Bioinformatics 10(1):41–48

Opazo JC, Homan FG, Storz JF (2008) Genomic evidence for independent origins of like globin genes in monotremes and therian mammals. Proc Natl Acad Sci U S A 105:1590–1595

Pace NR (1997) A molecular view of microbial diversity and the biosphere. Science 276:734–740

Padmavathi J, Uma Devi K, Rao CUM, Reddy NNR (2003) Telomere fingerprinting for assessing chromosome number, isolating typing and recombination in the entomopathogen *Beauveria bassiana*. Mycol Res 107(5):572–580

Page RDM (1998) GeneTree: comparing gene and species phylogenies using reconciled trees. Bioinformatics 14:819–820

Pagel M, Meade A (2004) A phylogenetic mixture model for detecting pattern-heterogeneity in gene sequence or character-state data. Syst Biol 53:571–581

Pérez-Losada M, Jobes DV, Sinangil F, Crandall KA, Posada D, Berman PW (2010) Phylodynamics of HIV-1 from a phase-III AIDS vaccine trial in North America. Mol Biol Evol 27:417–425

Plantier J-C, Leoz M, Dickerson JE, De Oliveira F, Cordonnier F, Lemee V, Damond F, Robertson DL, Simon F (2009) A new human immunodeficiency virus derived from gorillas. Nat Med 15:871–872

Posada D, Crandall KA, Templeton AR (2000) GeoDis: a program for the cladistic nested analysis of the geographical distribution of genetic haplotypes. Mol Ecol 9:487–488

Pozio E, Hoberg E, La Rosa G, Zarlenga DS (2009) Molecular taxonomy, phylogeny and biogeography of nematodes belonging to the *Trichinella genus*. Infect Genet Evol 9(4):606–616

Price MN, Dehal PS, Arkin AP (2009) FastTree: computing large minimum-evolution trees with profiles instead of a distance matrix. Mol Biol Evol 26:1641–1650

Ramírez-Flandes S, Ulloa O (2008) Bosque: integrated phylogenetic analysis software. Bioinformatics 24(21):2539–2541

Raphaël H, Milinkovitch MC (2010) MetaPIGA v2.0: maximum likelihood large phylogeny estimation using the metapopulation genetic algorithm and other stochastic heuristics. BMC Bioinforma 11:379

Rea PA, Kim Y, Sarafian V, Poole RJ, Davies JM, Sanders D (1992) Vacuolar H+-translocating pyrophosphatase: a new category of ion translocase. Trends Biochem Sci 17(9):348–352

Saitou N, Nei M (1987) The neighbor-joining method: a new method for reconstructing phylogenetic trees. Mol Biol Evol 4(4):406–425

Salemi M, Lamers SL, Yu S, de Oliveira T, Fitch WM, McGrath MS (2005) Phylodynamic analysis of human immunodeficiency virus type 1 in distinct brain compartments provides a model for the neuropathogenesis of AIDS. J Virol 79:11343–11352

Schmidt HA, Strimmer K, Vingron M, von Haeseler A (2002) TREE-PUZZLE: maximum likelihood phylogenetic analysis using quartets and parallel computing. Bioinformatics 18:502–504

Shimodaira H, Hasegawa M (1999) Multiple comparisons of log-likelihoods with applications to phylogenetic inference. Mol Biol Evol 16:1114–1116

Sokal RR, Michener CD (1958) A statistical method for evaluating systematic relationships. J Univ Kans Sci Bull 28:1409–1438

Suchard MA, Redelings BD (2006) BAli-Phy: simultaneous Bayesian inference of alignment and phylogeny. Bioinformatics 22:2047–2048

Suneetha G, Neelapu NRR, Surekha C (2016) Plant vacuolar proton pyrophosphatases (VPPases): structure, function and mode of action. Int J Recent Sci Res 7(6):12148–12152

Swofford DL (1991) PAUP: Phylogenetic Analysis Using Parsimony, version 3.1 Computer program distributed by the Illinois Natural History Survey, Champaign, Illinois

Swofford DL, Olsen GJ, Waddell PJ, Hillis DM (1996) Phylogenetic inference. In: Hillis DM, Moritz C, Mable BK (eds) Molecular systematics. Sinauer, Sunderland

Takahashi K, Nei M (2000) Efficiencies of fast algorithms of phylogenetic inference under the criteria of maximum parsimony, minimum evolution, and maximum likelihood when a large number of sequences are used. Mol Biol Evol 17:1251–1258

Teugels G (1996) Taxonomy, phylogeny and biogeography of catfishes (*Ostariophysi, Siluroidei*): an overview. Aquat Living Resour 9(S1):9–34. https://doi.org/10.1051/alr:1996039

Thompson RCA (2008) The taxonomy, phylogeny and transmission of *Echinococcus*. Exp Parasitol 119(4):439–446

Van Heuverswyn F, Peeters M (2007) The origins of HIV and implications for the global epidemic. Curr Infect Dis Rep 9:338–346

Vinh LS, von Haeseler A (2004) IQPNNI: moving fast through tree space and stopping in time. Mol Biol Evol 21(8):1565–1571

Woese CR (1987) Bacterial evolution. Microbiol Rev 51:221–271

Woese CR, Fox GE (1997) Phylogenetic structure of the prokaryotic domain: the primary kingdoms. Proc Natl Acad Sci U S A 74:5088–5090

Woese CR, Kandler O, Wheelis ML (1990) Towards a natural system of organisms: proposal for the domains Archaea, Bacteria, and Eucarya. Proc Natl Acad Sci U S A 87:4576–4579

Worobey M, Gemmel M, Teuwen DE, Haselkorn T, Kunstman K, Bunce M, Muyembe J-J, Kabongo J-MM, Kalengayi RM, Van Marck E, Gilbert MTP, Wolinsky SM (2008) Direct evidence of extensive diversity of HIV-1 in Kinshasa by 1960. Nature 455:661–664

Xia X, Xie Z (2001) DAMBE: data analysis in molecular biology and evolution. J Hered 92:371–373

Yang Z (2000) Phylogenetic analysis by maximum likelihood (PAML). University College, London

Chapter 11
Bioinformatics Insights on Plant Vacuolar Proton Pyrophosphatase: A Proton Pump Involved in Salt Tolerance

Nageswara Rao Reddy Neelapu, Sandeep Solmon Kusuma, Titash Dutta, and Challa Surekha

Contents

11.1   Introduction.. 193
11.2   Molecular Phylogeny of VPPase.. 195
11.3   Motifs of VPPase.. 196
11.4   Structure of VPPase.. 198
      11.4.1   Topology... 199
      11.4.2   Metal Geometry... 205
            11.4.2.1   Regulation of VPPase Enzyme Activity............................ 206
11.5   VPPase and Its Activity.. 207
11.6   Conclusion.. 208
References... 209

11.1   Introduction

Drought, salinity, and extreme temperatures are the major abiotic stress factors that adversely affect plant growth, development, and crop productivity. They alleviate the photosynthetic activity and induce nutrient scarcity and ionic and osmotic stress conditions in plants (Munns and Tester 2008; Rehman et al. 2005; Ashraf et al. 2008).

Salinity leads to degradation of soil fertility as a result of both natural and anthropogenic activities such as irrigation in arid and semiarid regions. Approximately 20% of the irrigated lands, i.e., 45 million hectares, is affected by soil salinization worldwide (Yeo 1999; Munns and Tester 2008). Moreover the change in global climate made rainfall less predictable and caused a drastic shift in the general rainfall pattern. This is of serious concern as there is much decrease in rainfed farm lands which produce one third of the world's food supply.

N. R. R. Neelapu · S. S. Kusuma · T. Dutta · C. Surekha (✉)
Department of Biochemistry and Bioinformatics, Institute of Science, Gandhi Institute of Technology and Management (GITAM) Deemed to be University, Visakhapatnam, AP, India
e-mail: surekha.challa@gitam.edu

© Springer Nature Switzerland AG 2019
K. R. Hakeem et al. (eds.), *Essentials of Bioinformatics, Volume III*,
https://doi.org/10.1007/978-3-030-19318-8_11

Under high salinity, plants experience both osmotic and ionic stress. The salt concentrations outside the roots rise rapidly, thereby leading to inhibition of water uptake by the roots, cell expansion, and lateral bud development (Munns and Tester 2008). Ionic stress develops when excess Na^+ accumulates particularly in leaves leading to increase in leaf mortality with chlorosis and necrosis and subsequently decrease in essential cellular metabolism activities such as photosynthesis (Yeo and Flowers 1986; Glenn et al. 1999). As NaCl is the most soluble and widespread salt, all plants have evolved mechanisms to regulate its accumulation.

Under salinity stress, plant cells need to maintain low cytosolic Na^+ level and high K^+ levels, resulting in a high cytosolic K^+/Na^+ ratio that is crucial for vital cellular metabolisms (Jeschke 1984; Blumwald 2000). The strategies generally employed by plants for the maintenance of a high K^+/Na^+ ratio in the cytosol include Na^+ extrusion and/or the intracellular compartmentalization of Na^+ (mainly in the plant vacuole). These mechanisms are vital for detoxification of cellular Na^+ levels and cellular osmotic adjustment which are needed to tolerate salt stress and plant survival (Blumwald 2000; Gaxiola et al. 2001; Li et al. 2010; Wei et al. 2011). The compartmentalization of Na^+ into vacuoles prevents the deleterious effects of Na^+ in the cytosol and allows the plants to use NaCl as an osmoticum. NaCl generates an osmotic potential that drives water into the cells (Gutiérrez-Luna et al. 2018).

The plant cell vacuole performs important biological functions such as recycling of cell components, regulation of turgor pressure, detoxification of xenobiotics, and accumulation of many useful substances. A large number of vacuolar proteins are known to be involved in support of the above multifaceted functions (Ohnishi et al. 2018). They include active pumps, carriers, ion channels, receptors, and structural proteins. Several major proteins of the tonoplast have been extensively investigated, and it was found that the three most abundant proteins of the tonoplast are vacuolar H^+-ATPase (V-ATPase), H^+-pyrophosphatase (V-PPase) (Maeshima 2000, 2001; Meng et al. 2017), and water channels (aquaporins) (King et al. 2004).

V-ATPase and VPPase coexist on the plant vacuolar membrane and use ATP and inorganic pyrophosphate (PPi), respectively, as energy sources for generating an electrochemical gradient of protons across the tonoplast. This facilitates the functioning of the Na^+/H^+-antiporter. The V-ATPase enzyme is a multisubunit proton pump found in all eukaryotes consisting of the peripheral (V1) complex responsible for ATP hydrolysis and the membrane-integral (Vo) complex responsible for proton translocation. V-ATPase is the largest complex in the tonoplast, with a total molecular size of about 750 kDa.

VPPase is a heat stable single polypeptide found in plants, algae, photosynthetic bacteria, protozoa, and archaebacteria (Rea et al. 1992; Maeshima 2000). It functions as a tonoplast proton pump and helps in Na^+ compartmentation. In plants, two isoforms of VPPase have been identified; one is potassium-dependent, while the other is potassium-independent (Belogurov and Lahti 2002; Schilling et al. 2017). Aquaporins are referred to as intercellular water channels imbedded in the membranes, and they facilitate transport of water, small solutes, and ions across membranes (Aharon et al. 2003; Porcel et al. 2005).

In this chapter, the vacuolar transporter VPPase has been reviewed with respect to its structure, function, phylogeny, and mode of action. This provides us with an understanding how plants tolerate and survive under salt-stressed environments.

11.2 Molecular Phylogeny of VPPase

VPPases have been reported to be highly conserved among land plants and less among archaeon, protozoan, and bacteria (Suneetha et al. 2016). VPPase from *R. rubrum* (Baltscheffsky et al. 1998), *Acetabularia acetabulum* (marine algae) (Ikeda et al. 1999), and *Chara coralline* (green algae) (Nakanishi et al. 1999) predicted the overall identities of amino acid sequences among these three phylogenetically separated organisms. It was reported that *R. rubrum* PPase synthase (660 residues) exhibited 36–39% with V-PPases of land plants and 40% with *A. acetabulum* V-PPase. Moreover *A. acetabulum* V-PPase shared 47% identity with land plant VPPases. However, the highest identity was observed in case of *C. corallina* (71%) with respect to land plants. These observations of sequence similarity suggest that *C. corallina* is evolutionarily closer to land plant than *R. rubrum* and *A. acetabulum*. Phylogeny with respect to other land plants revealed that VPPase of *A. thaliana* (AtVPP), *H. vulgare* (HvVPP), *B. vulgaris* (BvVPP), *N. tabacum* (NtVPP), and *O. sativa* (OsVPP) ranged from 761 to 771 amino acids in length. The amino acid sequences were found to be highly conserved with 86–91% sequence similarity among the land plants.

Phylogeny is used in establishing the origin and evolution pattern of a gene of particular species with respect to the other species. Generally phylogenetic tree is constructed using neighbor-joining (NJ) or maximum parsimony (MP) or maximum likelihood (ML) method (Saitou and Nei 1987). Suneetha et al. (2016) carried out phylogenetic studies on land plants, archaea, and bacterial V-PPases (Fig. 11.1).

Suneetha (2015) generated three phylogenetic trees in land plants using NJ, MP, and ML which showed similar topologies in both distance and character methods but differed in their branching order. Topological similarity of the trees obtained by different methods (NJ, MP, and ML) indicates that these clusters are not incidental and branching order reflected the expected pattern in all plants. The MP tree was constructed from 772 characters, out of which 515 were observed as conserved and 255 were variable, and of these 183 were parsimony informative. The tree length (L), consistency index (CI), and retention index (RI) in land plants were found to be 677, 0.61, and 0.77. The ML tree has a significant maximum likelihood tree length (−6594.00) (Fig. 11.2).

Similarly, Liu et al. (2011) reported that VPPase isolated from *Suaeda corniculata* showed highest similarity with *Kalidium foliatum* (96%), *Suaeda salsa* (94%), *Chenopodium rubrum* (89%), *Beta vulgaris* (89%), *Chenopodium glaucum* (88%), and *Arabidopsis thaliana* (87%). Dong et al. (2011) reported that apple VPPase (MdVHP1) shared highest similarity with peach VPPase (94%) followed by 87% similarity with VPPases of tobacco, grapevine, and *Arabidopsis*. Similarly, VPPase

Fig. 11.1 Relationship of 28 VPPases among land plants, land plant precursor, and bacteria as represented in a phylogenetic tree. (Source: Suneetha et al. 2016)

of *H. caspica* showed high sequence similarity with VPPases from Chenopodiaceae family and shared 95% sequence identity with VPPase of *K. foliadum*. All the studies reported the evolutionary history and relationship of VPPase gene among bacteria, land plants, and its precursor. The studies also provided enough evidence to conclude that VPPase gene is highly conserved among plant family members.

11.3   Motifs of VPPase

The structural model of VPPase showing N- and C-terminals in vacuolar end, transmembrane helices, and three conserved regions (CS1, CS2, and CS3) was reported by Maeshima (2001). Immunochemical analysis confirmed that these conserved sequences are located in the cytosolic loops (Takasu et al. 1997).

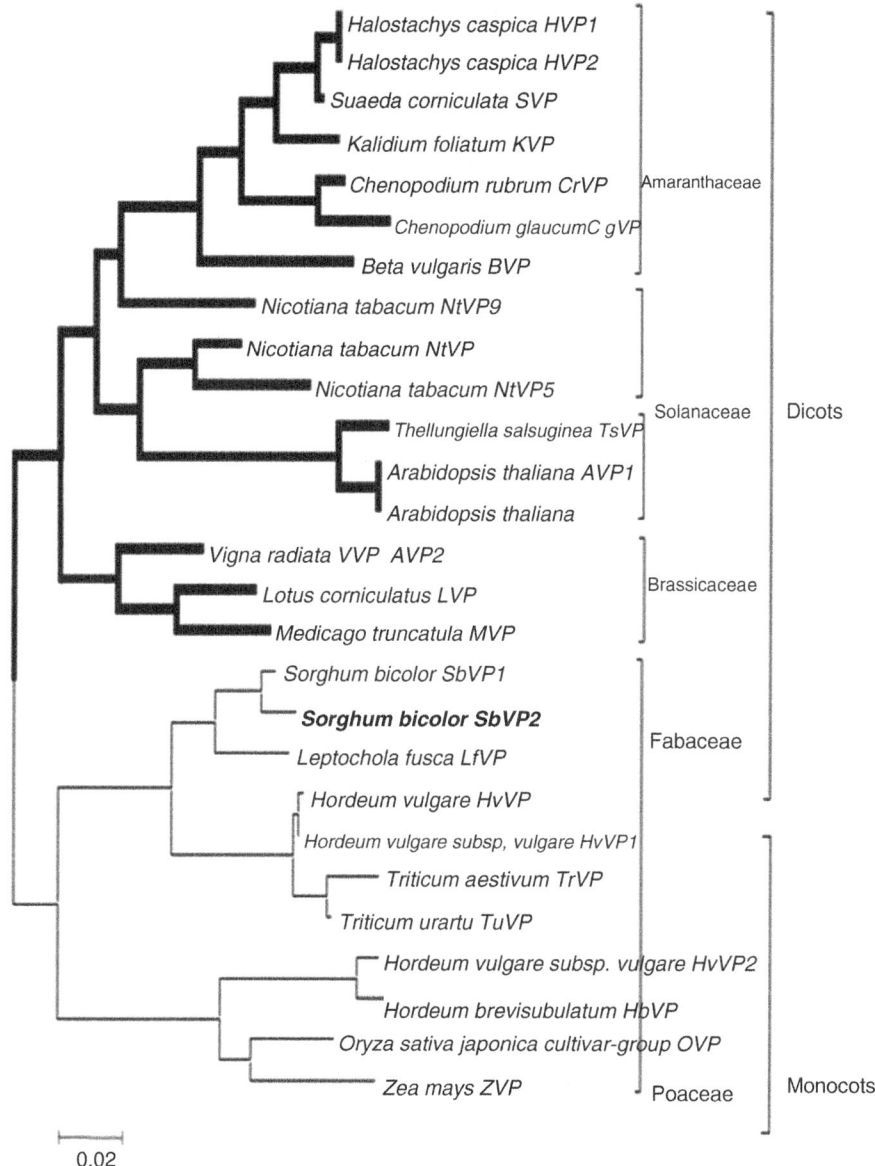

Fig. 11.2 Relationship of VPPases among land plants. The phylogenetic tree was generated using maximum likelihood method. (Source: Suneetha 2015)

Comparison of all VPPase genes from *C. coralline, A. acetabulum, R. rubrum,* and land plants reported with three highly conserved regions called motifs. The conserved motifs have been designated as CS1, CS2, and CS3 motifs (Rea and Poole 1993; Baltscheffsky et al. 1999; Maeshima 2000; Mimura et al. 2004; Suneetha 2015). Plant VPPase are characterized by the presence of cytosolic loops

(CLs), vacuolar loops (VLs), and transmembrane domains (TMDs) besides the N- and C-terminals residues (Zhen et al. 1997). Site-directed mutagenesis and immunochemical analysis revealed that the cytosolic domains are more conversed than the vacuolar domains and thus are crucial for VPPase enzyme activity.

The first conserved segment (CS1) has consensus sequence of DVGADLVGKVE and functions as the catalytic domain for substrate hydrolysis (Rea and Poole 1993; Schocke and Schink 1998). In addition to the catalytic site, there are binding sites for Mg^{2+}, K^+, and reagents, such as N,N-dicyclohexylcarbodiimide (DCCD), 7-chloro-4-nitrobenzo-2-oxa- 1,3-diazole (NBDCl), and N-ethylmaleimide (NEM) (Maeshima 2000; Sanders et al. 1999). Fukuda et al. (2004) validated the presence of NEM binding site at Cys-635 position, and Glu-306, Asp-505, and Glu-752 positions were identified as DCCD binding residues in barley. Zhen et al. (1997) conducted mutation and biochemical assays and revealed that Glu305 and Asp504 of A. thaliana V-PPase directly participate in DCCD binding and are presumably critical for catalysis.

The second conserved segment (CS2) is highly conserved and is located in a hydrophilic loop in the cytosol end. Suneetha (2015) reported that the CS2 motif has consensus sequence GSAALVSL and is approximately located at amino acid positions 543–550 in *Sorghum bicolor*. Suneetha (2015) reported that CS2 motif has function similar to rhodopsin like G-protein-coupled receptors (GPCRs) and is equipped with unique calcium signaling signature property that senses the high cytosolic Ca^{2+} levels and initiates V-PPase activity.

The third conserved segment (CS3) is located in the carboxyl-terminal part and contains 12 charged residues. It has consensus sequence GDTIGD exposed to the cytosol and plays a critical role in catalytic function in association with CS1 and CS2 segments (Liu et al. 2011; Rea et al. 1992). The position of these conserved regions change from one plant VPPase to others. For example, CS1 functional motifs DDPR and VGDN are located at 271 and 285 amino acid positions in mung bean, whereas in *S. corniculata* they are located at 266 and 280 amino acid positions, and in *S. bicolor* they occupy the 266 and 281 amino acid positions (Fig. 11.3). Similarly, the other conserved sequences CS2 and CS3 motifs are also highlighted in amino acid sequence alignment.

11.4 Structure of VPPase

Vacuolar H^+-pyrophosphatase (VPPase) catalyzes electrogenic H^+-translocation from the cytosol to the vacuolar lumen at the expense of hydrolysis of inorganic pyrophosphate (PPi). PPi is produced as a by-product of several metabolic processes, such as polymerization of DNA and RNA and synthesis of aminoacyl-tRNA (protein synthesis), ADP-glucose (starch synthesis), UDP-glucose (cellulose synthesis), and fatty acyl- CoA (L-oxidation of fatty acid).

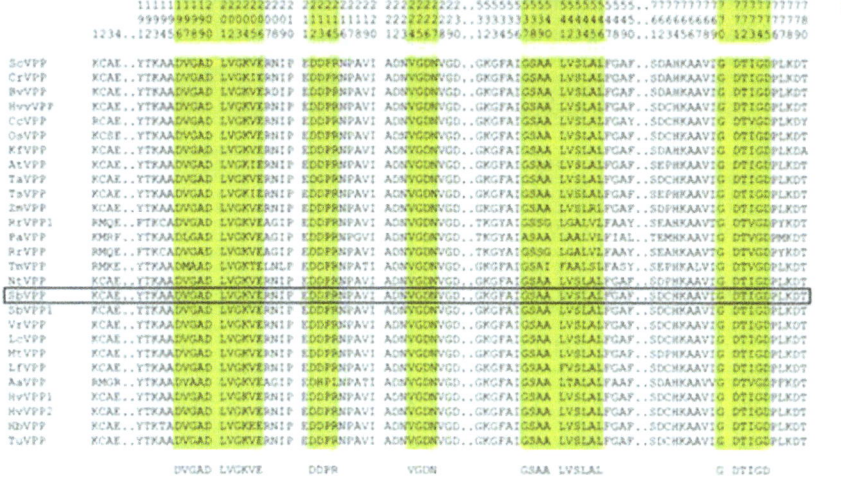

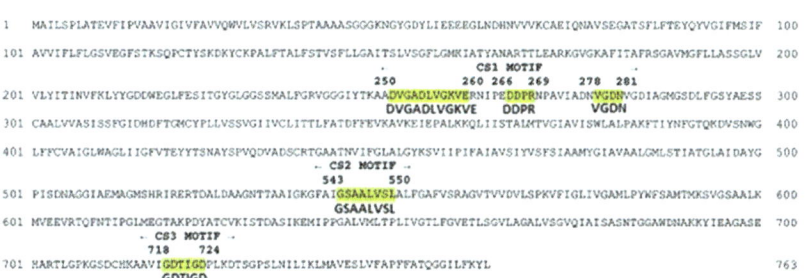

Fig. 11.3 Three conserved motifs CS1, CS2, and CS3 highlighted in (**a**) amino acid alignment are generated for sequences of VPPases; (**b**) region of conserved sequences of CS1, CS2, and CS3 are highlighted taking *S. bicolor* VPPase (meta-analysis of motifs was carried out)

11.4.1   Topology

VPPase consists of a single polypeptide, and its substrate, inorganic pyrophosphate (PPi), is one of the simplest high-energy compounds (Baltscheffsky et al. 1999; Maeshima 2000; Rea and Poole 1993). V-PPase gene encodes a polypeptide with 761–771 amino acids. Various V-PPase genes have been analyzed from different plant and bacterial species (Table 11.1). It was reported that VPPase gene isolated from *H. capsica* encodes 764 amino acids, apple VPPase gene encodes 771 amino acids, *S. corniculata* encodes 764, and *S. bicolor* encodes 763 amino acids.

Hydropathic and membrane topological analyses indicated that VPPase in general consists of 4–17 transmembrane domains (Table 11.2). Suneetha (2015) predicted that *S. bicolor* VPPase has 16 transmembrane regions using TMpred and

Table 11.1 List of VPPase genes and corresponding transmembrane helices

Sl no	Plant species	Accession ID	Transmembrane helices I→O	O→I	No. of amino acids	References
1	*Arabidopsis thaliana* L.	BAA32210	14	14	770	Sarafian et al. (1992)
		AAF31163	17	16	800	
		AAG09080	17	16	802	
2	*Beta vulgaris* L.	AAA61609	13	14	761	Kim et al. (1994a, b)
		AAA61610	14	15	765	
3	*Vigna radiata* L.	P21616	14	13	766	Nakanishi and Maeshima (1998)
4	*Cucurbita moschata*	BAA33149	14	14	768	Maruyama et al. (1998)
5	*Nicotiana tabacum* L.	Q43797	13	13	766	Lerchl et al. (1995)
		Q43798	13	15	765	
6	*Oryza sativa* L.	BAA08232	14	15	771	Sakakibara et al. (1996)
		BAA31524	15	16	767	
7	*Hordeum vulgare* L.	BAA02717	14	15	762	Tanaka et al. (1993)
8	*Triticum aestivum* L.	AAP55210	14	15	762	Brini et al. (2005)
9	*Cucumis sativus* L.	ABN48304	3	3	161	Kabala et al. (2008)
10	*Vitis vinifera* L.	NP001268155	14	14	764	Da Silva et al. (2013)
		CAD89675	14	14	764	Venter et al. (2006)
11	*Medicago truncatula*	AES91661	12	14	624	Young et al. (2011)
		AES91660	13	14	765	
		KEH28512	17	17	799	
12	*Triticum urartu*	EMS67279	9	9	414	Ling et al. (2013)
		EMS65629	14	15	762	
		EMS53286	14	14	700	
13	*Zea mays* L.	NP001168714	1	1	97	Schnable et al. (2009)
		AFW77254	13	15	765	
		AFW70478	13	15	766	
14	*Salicornia europaea* L.	AEI17666	14	15	763	Lv et al. (2012)
		AEI17665	14	14	764	
15	*Aeluropus littoralis*	ALO51665	14	16	763	Ebrahimi et al. (2015)
16	*Ipomoea babatas* L.	AFQ00710	13	14	767	Fan (2011)
17	*Oxybasis rubra* L.	AAM97920	14	14	764	Kranewitter et al. (2002)
18	*Sorghum bicolor* L.	ACV74424	15	16	763	Anjaneyulu et al. (2014)

(continued)

Table 11.1 (continued)

Sl no	Plant species	Accession ID	Transmembrane helices I→O	O→I	No. of amino acids	References
19	*Solanum lycopersium* L.	NP_001307479	14	15	767	Mohammed et al. (2012)
		BAM65603	13	13	765	
		BAM65604	17	16	800	
20	*Kalidium foliatum*	ABK91685	14	14	764	Yao et al. (2012)
21	*Chara corallina*	AB018529	15	16	793	Nakanishi et al. (1999)
22	*Acetabularia acetabulum*	D88820	13	14	721	Ikeda et al. (1999)
23	*Rhodospirillum rubrum*	AF044912	14	15	660	Baltscheffsky et al. (1998)

Table 11.2 Sequence positions of possible transmembrane helices from inside to outside and vice versa in VPPase of *S. bicolor*

Inside to outside helices				Outside to inside helices			
From	To	Score	Centre	From	To	Score	Centre
11(14)	31 (28)	2024	221	11 (14)	31 (28)	2589	21
95 (95)	111 (111)	2670	104	92 (92)	111 (108)	2664	101
136 (138)	156 (153)	2054	146	136 (136)	156 (156)	2109	146
188 (188)	206 (204)	2150	197	188 (191)	206 (206)	2147	198
229 (229)	244 (244)	682	236	223 (227)	244 (242)	860	234
296 (296)	314 (310)	1049	303	293 (293)	314 (309)	1058	302
325 (325)	342 (342)	2139	335	326 (326)	342 (342)	2143	333
366 (366)	383 (383)	1886	373	361 (361)	381 (379)	1974	372
401 (401)	417 (415)	2725	408	401 (402)	418 (418)	2629	409
456 (460)	476 (476)	2120	467	456 (456)	475 (475)	2120	466
476 (476)	494 (490)	1990	483	471 (473)	490 (488)	1997	481
539 (541)	557 (557)	2092	550	539 (542)	557 (557)	1991	549
573 (573)	589 (589)	1938	581	572 (572)	588 (588)	1672	581
644 (644)	659 (659)	1324	652	639 (642)	659 (657)	1609	649
661 (661)	681 (678)	1216	669	661 (664)	680 (678)	1256	671
740 (743)	760 (757)	1314	750	740 (740)	760 (757)	1238	750

TMHMM. The results obtained showed that the sequences has 16 inside to outside helices orientations and 16 outside to inside helices orientations of the transmembranes (Fig. 11.4a, b).

The 3D structure of VPPase is a vacuolar membrane-bound protein compactly folded in rosette manner in two concentric walls (Lin et al. 2012; Suneetha et al. 2016; Suneetha 2015) (Fig. 11.5). Lin et al. (2012) reported that mung bean VPPase has 16 transmembrane helices, but it exists as a homodimer, and Suneetha (2015)

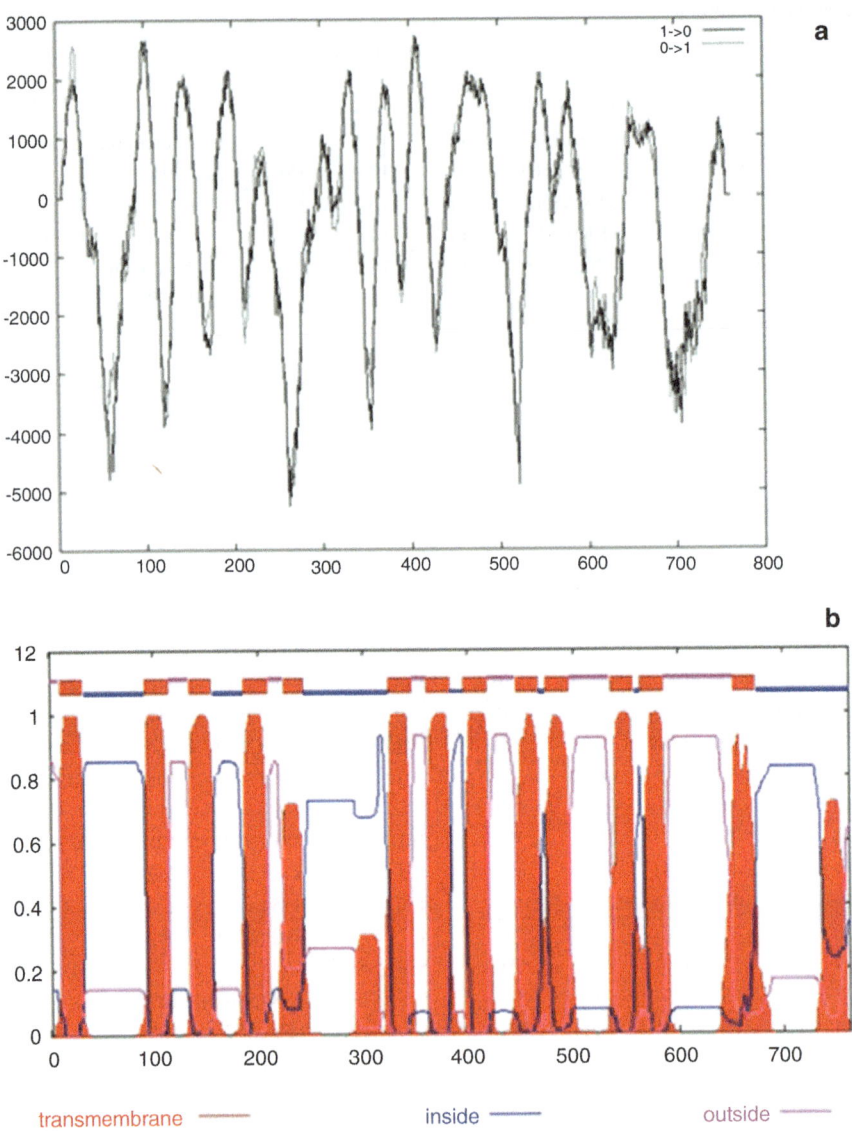

Fig. 11.4 Transmembrane helices of VPPase in *S. bicolor* (**a**) TM pred, (**b**) TMHMM. (Source: Suneetha 2015)

reported that *S. bicolor* VPPase exists as monomer with 16 transmembrane helices. The core has six transmembrane helices surrounded by ten transmembrane helices which form the inner and outer walls of the pump which is displayed in cylinders (Fig. 11.6). Two short helices are present on the cytosolic side; two helices and two antiparallel β-strands are present on the luminal side of the protein (Fig. 11.7).

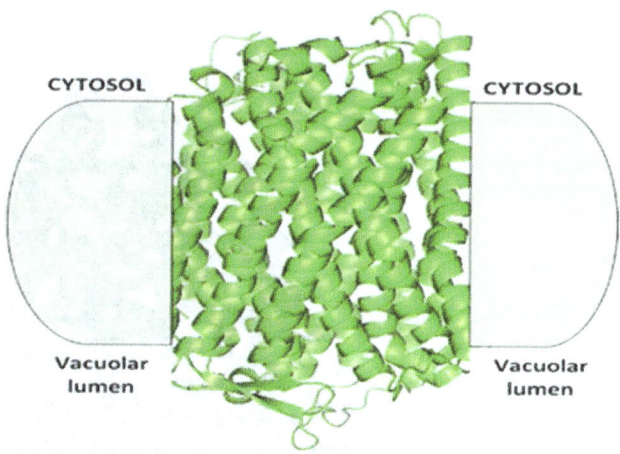

Fig. 11.5 VPPase protein compactly folded as membrane-bound protein. (Source: Suneetha et al. 2016)

Fig. 11.6 Sixteen transmembrane helices (blue cylinders) with six helices in the core surrounded by ten transmembrane helices to form inner and outer walls of the pump. (Source: Suneetha 2015)

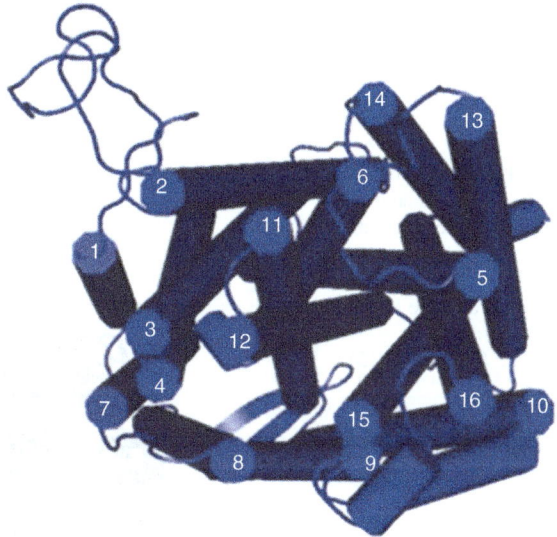

The core of the model has one IDP molecule surrounded by five Mg^{2+} ions which are essential for the activity of V-PPases and one K^+ ion which acts as stimulator (Fig. 11.8). The above elements are highly conserved among the VPPases which forms a hydrophobic door to the hydrophilic surroundings of the vacuolar lumen. The hydrophobic gate prevents the reflux of H^+ ions and helps in maintaining the translocation of H^+ from cytosol to vacuolar lumen (Fig. 11.9). The space-fill representation of VPPase model is considered to analyze electrostatic surface potential.

Fig. 11.7 Ribbon structure of VPPase containing 16 transmembrane helices (colored in blue) and antiparallel β-strands (colored in red). (Source: Suneetha 2015)

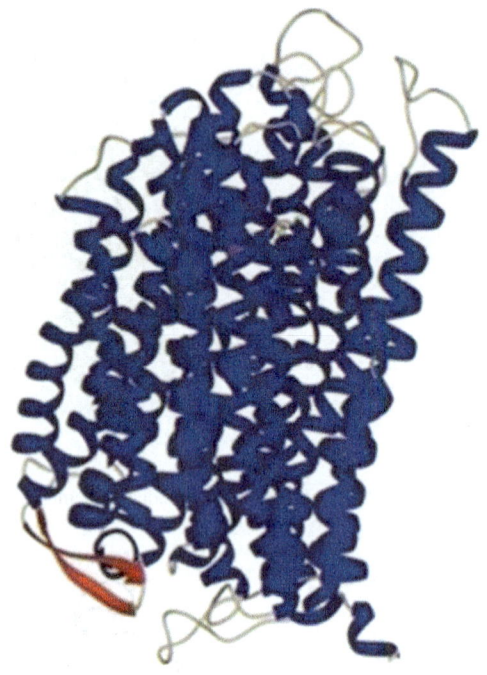

Fig. 11.8 VPPase model of *S. bicolor* rotated to 600 to visualize the core with one imidodiphosphate (IDP), five Mg2+ (colored in green), and one K+ ions (colored in purple). (Source: Suneetha 2015)

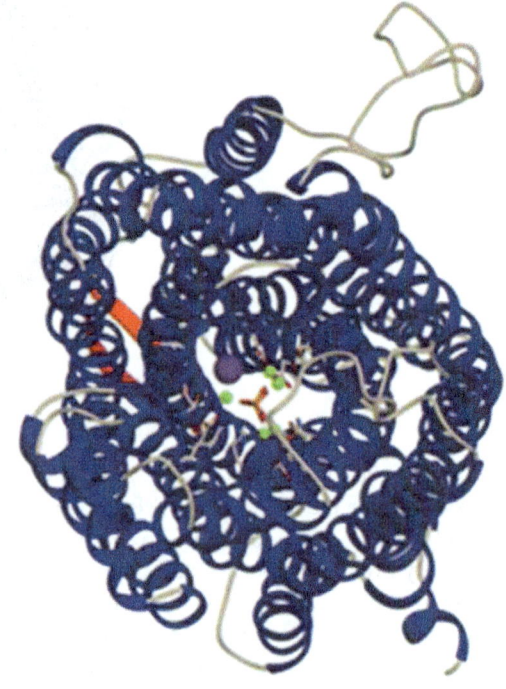

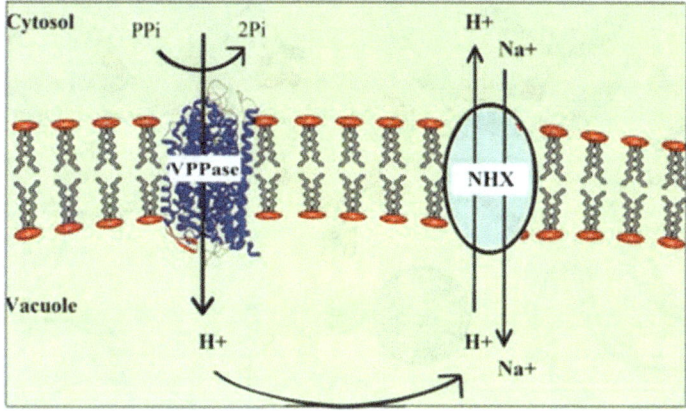

Fig. 11.9 Working model of the VPPase showing the pumping of protons into vacuole to generate electrochemical gradient against which sodium is taken in under stress conditions. (Source: Suneetha 2015)

Fig. 11.10 The space-fill representation of modeled VPPase showing electrostatic surface potential. The electrostatic surface negative potential (red), positive (blue), and neutral (white) are represented. The core of the model contains IDP binding site. (Source: Suneetha 2015)

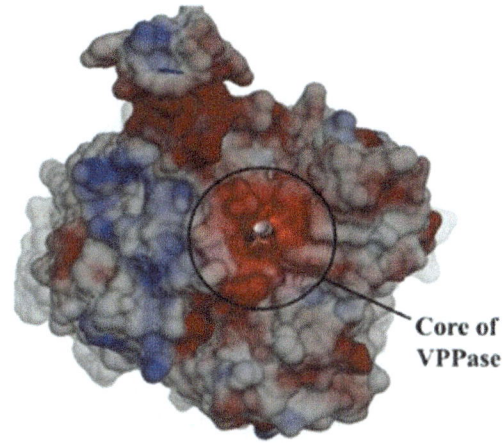

Core of VPPase

The surface potential is indicated by colors as in Fig. 11.10. The core of model which contains IDP binding site is represented within the circle the core of VPPase (Fig. 11.10).

11.4.2 Metal Geometry

V-PPase requires free Mg^{2+} as an essential cofactor. $MgCl_2$ and $MgSO_4$ are added to the buffers for solubilization and purification of the enzyme during its isolation (Maeshima and Yoshida 1989; Britten et al. 1989; Rea and Poole 1986). Binding of

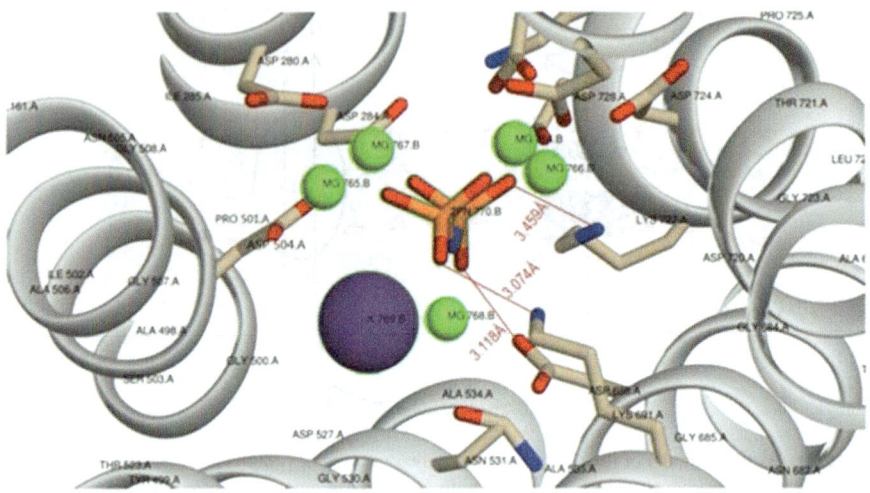

Fig. 11.11 The core of VPPase showing coordinating amino acids from IDP molecule to five Mg2+ ions. (Source: Suneetha 2015)

Mg^{2+} stabilizes and activates the enzyme. Baykov et al. (1993) reported the presence of high-affinity and low-affinity Mg^{2+} binding sites of mung bean. Binding of Mg^{2+} to VPPase not only activates the enzyme but also protects it from heat inactivation (Baykov et al. 1993). Suneetha (2015) reported that the core has five Mg^{2+} and one K^+ ions along with one IDP which play an important role in activating VPPases by transphosphorylation reaction involving ATP's. Each Mg^{2+} ion interacts with surrounding amino acids like aspartic acid, asparagines, and glutamic acid (Fig. 11.11). Potassium ion acts as stimulator of VPPase and is surrounded by amino acids like asparagine and glycine. K^+ stimulates VPPase activity by more than threefold in most cases (Gordon-Weeks et al. 1999). The maximal activity of VPPase was obtained in the presence of more than 30 mM KCl in most cases. Suneetha (2015) also reported that there are eleven phosphate binding sites represented in yellow color balls and interacting residues with green color (Fig. 11.12).

11.4.2.1 Regulation of VPPase Enzyme Activity

Studies on VPPase from various plant species revealed the relationship between the enzyme activity of the proton pump with respect to varying concentrations of cytosolic ions and chemical compounds. K^+ ions have been associated with increased VPPase enzyme activity in *A.thaliana* type 2 VPPase (AVP2). Ca^{2+} reversibly inhibits VPPase activity through formation of Ca-PPi which is a strong, competitive inhibitor for the soluble PPases (Baykov et al. 1999). Changes in free cytosolic Ca^{2+} levels have also been associated with negative inhibition of VPPase activity in bean guard cells (Darley et al. 1998) and barley (Swanson and Jones 1996).

Fig. 11.12 Eleven phosphate binding sites of VPPase are represented in yellow colored balls and interacting residues in green color. (Source: Suneetha 2015)

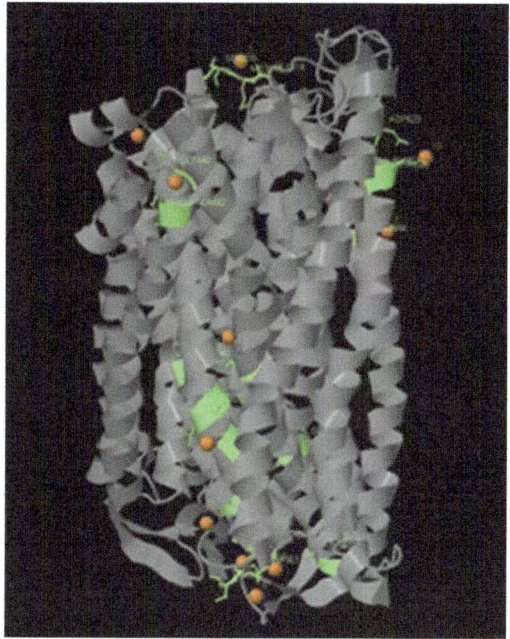

Cytosolic Mg^{2+} concentration has also been reported for optimum enzyme activity in *S. bicolor*, mung bean, and barley. Moreover excessive Na^+ concentrations have been reported to inhibit enzyme activity in red beet (Rea and Poole 1985).

Among the artificial substances tested, it reported that amino methylene bisphosphonate (AMBP) is a potent inhibitor of VPPase in mung bean and *A. thaliana* VPPase AVP2 and AVP1 (Zhen et al. 1994). The effectiveness of bisphosphonates as an inhibitor of VPPase was carried out, and it was concluded that a nitrogen atom in the carbon chain of bisphosphonates increased the inhibitory effect of the enzyme (Gordon-Weeks et al. 1999).

11.5 VPPase and Its Activity

Proton pump VPPase gets activated upon signals perceived by plants. The sequences of events occurring during the activation of proton pump are as follows:

Abiotic stress (high salinity, drought, high temperatures, etc.) in plants is perceived by root tissues and cells. The cells activate receptor-bound G-proteins to activate protein kinases by the breakdown of membrane-bound phosphatidylinositol bisphosphate (PIP2) to diacylglycerol (DAG) and inositol triphosphate (IP3) (Mahajan and Tuteja 2005; Tuteja 2007). IP3 induces endoplasmic reticulum in release of Ca^{2+} and other side; it also makes calcium channels to open to increase intracellular Ca^{2+} levels. CS1 and CS3 motifs form the core catalytic domain and are

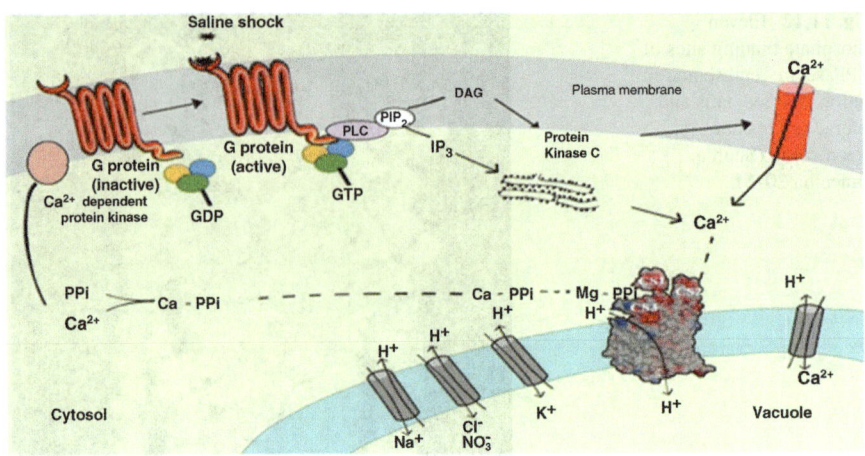

Fig. 11.13 Web of events which show saline shock initiating cascade of signals to generate PMF that drives sodium into vacuole leading to salt tolerance in plants

essential for hydrolyzing PPi and transport protons (Fig. 11.13). CS2 motif of VPPase, similar to rhodopsin-like G-protein-coupled receptor (GPCR) with calcium signaling signature property, senses these high cytosolic Ca^{2+} levels and transduces extracellular signal. The free available cytosolic Ca^{2+} may be phosphorylated to Ca-PPi by Ca^{2+}-dependent membrane-bound protein kinase and PPi (Johannsen et al. 1996). The substrate PPi of Ca-PPi is exchanged with Mg^{2+} to form Mg-PPi at the core catalytic site from CS1 and CS3.

The above elements are highly conserved among the VPPases that form hydrophobic door to the hydrophilic surroundings of vacuolar lumen. The acidic residues in the core catalytic site help in PPi hydrolysis and proton transport into vacuole. The hydrophobic gate prevents the reflux of H^+ ions and helps in maintaining the translocation of H+ from cytosol to vacuolar lumen. The pumping of H^+ into vacuole builds electrochemical gradient (proton motive force, PMF) which changes its pH (2–4 pH units, equivalent to −120 to −240 mV) (Isayenkov et al. 2010). The PMF can energize various antiporters such as Na^+ and K^+: H^+ exchanger, NO_3^- and Cl^-: H^+ exchanger, etc. resulting in influx of Na^+, K^+, NO_3^-, and Cl^- from cytosol to vacuole. This influx reduces the toxicity of cytosol to protect the cell against deleterious effects thus caused due to abiotic stress. Therefore, overall signaling web plays an important role in providing stress tolerance to plants.

11.6 Conclusion

Vacuolar transporters are vital components of cellular network. They enable the plant to respond to the changing environmental conditions, store nutrients and energy during surplus production, and maintain optimal metabolic conditions in the cytosol. Plant vacuolar VPPase, a model of proton pump, is considered as integral enzyme due to its structure-function relationship.

Structural analysis using both laboratory and bioinformatic approaches revealed the functional domains along with the conserved segments (CS1, CS2, and CS3) that play an active role in the translocation of H^+ ions into the vacuole from the cytosol. Phylogenetic analysis of all known VPPase across land plants, archaea, protozoan, and bacteria increased our knowledge of the tonoplast dramatically over the past decade. Studies established that during evolution of organisms, ancestral plant species obtained VPPase in addition to vacuolar-type V-ATPase.

However, more information is required on protein-ligand interactions and the molecular evolution of VPPase. It has been reported that the expression levels of VPPase change according to the physiological conditions and in response to environmental stresses. However, the regulatory mechanism and the posttranslational regulations of VPPase are yet to be studied. Thus, these analyses are extremely important toward establishing the role of VPPase as effective proton pump dedicated toward alleviating salt stress. The VPPase gene has been successfully used to engineer transgenic plants. Overexpression of the VPPase gene was able to confer effective Na^+ compartmentation into the vacuole. Moreover, various VPPase from other species can be isolated to study their functional properties and development of transgenic plants. Thus enabling the plant to survive during salt stress and maintain an optimum osmoticum of the cytosol.

Acknowledgments The authors are grateful to Gandhi Institute of Technology and Management (GITAM) deemed-to-be-university, for providing necessary facilities to carry out the research work and for extending constant support in writing this review.

References

Aharon R, Shahak Y, Wininger S, Bendov R, Kapulnik Y, Galili G (2003) Overexpression of a plasma membrane aquaporin in transgenic tobacco improves plant vigor under favorable growth conditions but not under drought or salt stress. Plant Cell 15:439–447

Anjaneyulu E, Reddy PS, Sunita MS, Kishor PBK, Meriga B (2014) Salt tolerance and activity of antioxidative enzymes of transgenic finger millet overexpressing a vacuolar H+-pyrophosphatase gene (SbVPPase) from *Sorghum bicolor*. J Plant Physiol 171(10):789–798

Ashraf M, Athar HR, Harris PJC, Kwon TR (2008) Some prospective strategies for improving crop salt tolerance. Adv Agron 97:45–110

Baltscheffsky M, Nadanaciva S, Schultz A (1998) A pyrophosphate synthase gene: molecular cloning and sequencing of the cDNA encoding the inorganic pyrophosphate synthase from *Rhodospirillum rubrum*. Biochim Biophys Acta 1364:301–306

Baltscheffsky M, Schultz A, Baltscheffsky H (1999) HC-PPases: a tightly membranebound family. FEBS Lett 457:527–533

Baykov AA, Bakuleva NP, Rea PA (1993) Steady-state kinetics of substrate hydrolysis by vacuolar H+-pyrophosphatase: a simple three-state model. Eur J Biochem 217:755–762

Baykov AA, Cooperman BS, Goldman A, Lahti R (1999) Prog Mol Subcell Biol 23:127–150

Belogurov GA, Lahti R (2002) A lysine substitute for K+-A460K mutation eliminates K+ dependence in H+-pyrophosphatase of Carboxydothermus hydrogenoformans. J Biol Chem 277:49651–49654

Blumwald E (2000) Sodium transport and salt tolerance in plants. Curr Opin Cell Biol 12:431–434

Brini F, Gaxiola RA, Berkowitz GA, Masmoudi K (2005) Cloning and characterization of a wheat vacuo-lar cation/proton antiporter and pyrophosphatase proton pump. Plant Physiol Biochem 43(4):347–354

Britten CJ, Turner JC, Rea PA (1989) Identification and purification of substrate-binding subunit of higher plant H+-translocating inorganic pyrophosphatase. FEBS Lett 256:200–206

Da Silva C, Zamperin G, Ferrarini A, Minio A, Dal Molin A, Venturini L, Buson G, Tononi P, Avanzato C, Zago E, Boido E (2013) The high polyphenol content of grapevine cultivar tannat berries is conferred primarily by genes that are not shared with the reference genome. Plant Cell 25(12):4777–4788

Darley CP, Skiera LA, Northrop FD, Sanders D, Davies JM (1998) Tonoplast inorganic pyrophos-phatase in Vicia faba guard cells. Planta 206:272–2777

Dong QL, Liu DD, An XH, Hu DG, Yao YX, Hao YJ (2011) MdVHP1 encodes an apple vacuolar H+-PPase and enhances stress tolerance in transgenic apple callus and tomato. J Plant Physiol 168(17):2124–2133

Ebrahimi A, Monfared SRA, Kashkooli AB (2015) Pyrophosphate-energized vacuolar membrane proton pump [*Aeluropus littoralis*] agronomy and plant breeding, Tehran University, Karaj, Alborz 31587-1167, Iran

Fan W (2011) Overexpression of the Na+/H+ antiporter gene from sweet potato. Cassava and sweetpotato biotechnology, direct submission to NCBI with accession no. AFQ00710

Fukuda A, Chiba K, Maeda M, Nakamura A, Maeshima M, Tanaka Y (2004) Effect of salt and osmotic stresses on the expression of genes for the vacuolar H+-pyrophosphatase, H+-ATPase subunit A, and Na+/H+ antiporter from barley. J Exp Bot 55(397):585–594

Gaxiola RA, Li J, Undurraga S, Dang LM, Allen GJ, Alper SL, Fink GR (2001) Drought-and salt-tolerant plants result from overexpression of the *AVP1* H+-pump. Proc Natl Acad Sci U S A 98(20):11444–11449

Glenn EP, Brown JJ, Blumwald E (1999) Salt tolerance and crop potential of halophytes. Crit Rev Plant Sci 18:227–255

Gordon-Weeks R, Parmar S, Davies TGE, Leigh RA (1999) Structural aspects of the effectiveness of bisphosphonates as competitive inhibitors of the plant vacuolar proton-pumping pyrophos-phatase. Biochem J 337:373–377

Gutiérrez-Luna FM, Hernández-Domínguez EE, Valencia-Turcotte LG, Rodríguez-Sotres R (2018) Pyrophosphate and pyrophosphatases in plants, their involvement in stress responses and their possible relationship to secondary metabolism. Plant Sci 267:11. https://doi.org/10.1016/j.plantsci.2017.10.016. Epub 2017 Nov 8.

Ikeda M, Tanabe E, Rahman MH, Kadowaki H, Moritani C et al (1999) A vacuolar inorganic HC-pyrophosphatase in *Acetabularia acetabulum*: partial purification, characterization and molecular cloning. J Exp Bot 50:139–140

Isayenkov S, Isner JC, Maathuis FJM (2010) Vacuolar ion channels: roles in plant nutrition and signalling. FEBS Lett 584:1982–1988

Jeschke WD (1984) K+-Na+ exchange at cellular membranes, intracellular compartmentation of cations, and salt tolerance. Sanity tolerance in plant. Strategies for crop improvement. Wiley-Interscience Publication, New York, pp 33–76

Johansson I, Larsson C, Ek B, Kjellbom P (1996) The major integral proteins of spinach leaf plasma membranes are putative aquaporins and are phosphorylated in response to Ca+ and apoplastic water potential. Plant Cell 8:1181–1191

Kim Y, Kim EJ, Rea PA (1994a) Isolation and characterization of cDNAs encoding the vacuolar HC-pyrophosphatase of *Beta vulgaris*. Plant Physiol 106:375–382

Kim Y, Kim EJ, Rea PA (1994b) Isolation and characterization of cDNAs encoding the vacuolar H+-pyrophosphatase of *Beta vulgaris*. Plant Physiol 106(1):375–382

King LS, Kozono D, Agre P (2004) From structure to disease: the evolving tale of aquaporin biol-ogy. Nat Rev Mol Cell Biol 5:678–698

Kranewitter W, Gogarten P, Pfeiffer W (2002) Cloning and sequencing of the vacuolar proton-pumping PPase from *Chenopodium rubrum*. Direct submission to NCBI with accession no. AAM97920

Lerchl J, K"onig S, Zrenner R, Sonnewald U (1995) Molecular cloning, characterization and expression analysis of isoforms encoding tonoplast-bound protontranslocating inorganic pyrophosphatase in tobacco. Plant Mol Biol 29:833–840

Li Z, Baldwin CM, Hu Q, Liu HB, Luo H (2010) Heterologous expression of *Arabidopsis* H+-pyrophosphatase enhances salt tolerance in transgenic creeping bentgrass (*Agrostis stolonifera* L.). Plant Cell Environ 33(2):272–289

Lin CH, Peng PH, Ko CY, Markhart AH, Lin TY (2012) Characterization of a novel Y2 K-type dehydrin *VrDhn1* from *Vigna radiata*. Plant Cell Physiol 53:930–942

Ling HQ, Zhao S, Liu D, Wang J, Sun H, Zhang C, Fan H, Li D, Dong L, Tao Y, Gao C (2013) Draft genome of the wheat A-genome progenitor *Triticum urartu*. Nature 496(7443):87

Liu L, Wang Y, Wang N, Dong YY, Fan XD, Liu XM, Li HY (2011) Cloning of a vacuolar H+-pyrophosphatase gene from the halophyte *Suaeda corniculata* whose heterologous overexpression improves salt, saline-alkali and drought tolerance in *Arabidopsis*. J Integr Plant Biol 53(9):731–742

Lv S, Jiang P, Chen X, Fan P, Wang X, Li Y (2012) Multiple compartmentalization of sodium conferred salt tolerance in *Salicornia europaea*. Plant Physiol Biochem 51:47–52

Maeshima M (2000) Vacuolar H+-pyrophosphatase. Biochim Biophys Acta 1465:37–51

Maeshima M (2001) Tonoplast transporters: organization and function. Annu Rev Plant Physiol Plant Mol Biol 52:469–497

Maeshima M, Yoshida S (1989) Purification and properties of vacuolar membrane protontranslocating inorganic pyrophosphatase from mung bean. J Biol Chem 264:20068–20073

Mahajan S, Tuteja N (2005) Cold, salinity and drought stresses: An overview. Arch Biochem Biophys 444:139–158

Maruyama C, Tanaka Y, Mitsuda NT, Takeyasu K, Yoshida M, Sato MH (1998) Structural studies of the vacuolar H+-pyrophosphatase: sequence analysis and identification of the residues modified by fluorescent cyclohexylcarbodiimide and maleimide. Plant Cell Physiol 39:1045–1053

Meng L, Li S, Guo J, Guo Q, Mao P, Tian X (2017) Molecular cloning and functional characterisation of an H+-pyrophosphatase from *Iris lactea*. Sci Rep 7(1):17779

Mimura H, Nakanishi Y, Hirono M, Maeshima M (2004) Membrane topology of the H+-pyrophosphatase of *Streptomyces coelicolor* determined by cysteine-scanning mutagenesis. J Biol Chem 279(33):35106–35112

Mohammed SA, Nishio S, Takahashi H, Shiratake K, Ikeda H, Kanahama K, Kanayama Y (2012) Role of vacuolar H+-inorganic pyrophosphatase in tomato fruit development. J Exp Bot 63(15):5613–5621

Munns R, Tester M (2008) Mechanisms of salinity tolerance. Annu Rev Plant Biol 59:651–681

Nakanishi Y, Maeshima M (1998) Molecular cloning of vacuolar H+-pyrophosphatase and its developmental expression in growing hypocotyl of mung bean. Plant Physiol 116:589–597

Nakanishi Y, Matsuda N, Aizawa K, Kashiyama T, Yamamoto K et al (1999) Molecular cloning of the cDNA for vacuolar H+-pyrophosphatase from *Chara corallina*. Biochem Biophys Acta 1418:245–250

Ohnishi M, Yoshida K, Mimura T (2018) Analyzing the vacuolar membrane (tonoplast) proteome. In: Plant membrane proteomics. Humana Press, New York, pp 107–116

Porcel R, Gomez M, Kaldenhoff R, Ruiz-Lozano JM (2005) Impairment of NtAQP1 gene expression in tobacco plants does not affect root colonisation pattern by arbuscular *mycorrhizal* fungi but decreases their symbiotic efficiency under drought. Mycorrhiza 15:417–423

Rea PA, Poole RJ (1985) Proton-translocating inorganic pyrophosphatase in red beet (Beta vulgaris L.) tonoplast vesicles. Plant Physiol 77:46–52

Rea PP, Poole RJ (1986) Chromatographic resolution of H+-translocating pyrophosphatase from H+-translocating ATPase of higher plant tonoplast. Plant Physiol 81:126–129

Rea PA, Poole RJ (1993) Vacuolar H+ −translocating pyrophosphatase. Annu Rev Plant Physiol Plant Mol Biol 44:157–180

Rea PA, Kim Y, Sarafian V, Poole RJ, Davies JM, Sanders D (1992) Vacuolar H+-translocating pyrophosphatase: a new category of ion translocase. Trends Biochem Sci 17(9):348–352

Rehman S, Harris PJC, Ashraf M (2005) Stress environments and their impact on crop production, Abiotic stresses: plant resistance through breeding and molecular approaches. Haworth Press, New York, pp 3–18

Saitou N, Nei M (1987) The neighbor-joining method: a new method for reconstructing phylogenetic trees. Mol Biol Evol 4:406–425

Sakakibara Y, Kobayashi H, Kasamo K (1996) Isolation and characterization of cDNAs encoding vacuolar H⁺-pyrophosphates isoforms from rice (*Oryza sativa* L.). Plant Mol Biol 31:1029–1038

Sanders D, Brownlee C, Harper JF (1999) Communicating with calcium. Plant Cell 11:691–706

Sarafian V, Kim Y, Poole RJ, Rea PA (1992) Molecular cloning and sequence of cDNA encoding the pyrophosphate-energized vacuolar membrane proton pump of *Arabidopsis thaliana*. Proc Natl Acad Sci 89(5):1775–1779

Schilling RK, Tester M, Marschner P, Plett DC, Roy SJ (2017) AVP1: one protein, many roles. Trends Plant Sci 22(2):154–162

Schnable PS, Ware D, Fulton RS, Stein JC, Wei F, Pasternak S, Liang C, Zhang J, Fulton L, Graves TA, Minx P (2009) The B73 maize genome: complexity, diversity, and dynamics. Science 326(5956):1112–1115

Schocke L, Schink B (1998) Membrane-bound proton-translocating pyrophosphatase of syntrophus gentianae, a syntrophically benzoate-degrading fermenting bacterium. Eur J Biochem 256:589–594

Suneetha G (2015) Studies on *in vitro*, *in planta* and *in silico* analysis of vacuolar proton pyrophosphatase from *Sorghum bicolor* (*Sb*V-PPase) and its overexpression in *Cajanus cajan* (unpublished doctoral thesis). GITAM University, Visakhapatnam, Andhra Pradesh, India

Suneetha G, Neelapu NRR, Surekha CH (2016) Plant vacuolar proton pyrophosphatases (VPPases): structure, function and mode of action. Int J Recent Sci Res Res 7(6):12148–12152

Swanson SJ, Jones RL (1996) Gibberellic acid induces vacuolar acidification in barley aleurone. Plant Cell 8:2211–2221

Takasu A, Nakanishi Y, Yamauchi T, Maeshima M (1997) Analysis of the substrate binding site and carboxyl terminal region of vacuolar H⁺-pyrophosphatase of mung bean with peptide antibodies. J Biochem 122:883–889

Tanaka Y, Chiba K, Maeda M, Maeshima M (1993) Molecular cloning of cDNA for vacuolar membrane proton-translocating inorganic pyrophosphatase in *Hordeum vulgare*. Biochem Biophys Res Commun 190:1110–1114

Tuteja N (2007) Mechanisms of high salinity tolerance in plants. Methods Enzymol 428:419–438

Venter M, Groenewald JH, Botha FC (2006) Sequence analysis and transcriptional profiling of two vacuolar H⁺-pyrophosphatase isoforms in *Vitis vinifera*. J Plant Res 119(5):469–478

Wei Q, Guo YJ, Cao H, Kuai BK (2011) Cloning and characterization of an *AtNHX2*-like Na⁺/H⁺ antiporter gene from *Ammopiptanthus mongolicus* (Leguminosae) and its ectopic expression enhanced drought and salt tolerance in *Arabidopsis thaliana*. Plant Cell Tissue Organ Cult 105(3):309–316

Yao M, Zeng Y, Liu L, Huang Y, Zhanq F (2012) Overexpression of the halophyte *Kalidium Foliatum* H⁺-pyrophosphatase gene confers salt and drought tolerance in *Arabidopsis thaliana*. Mol Biol Rep 39(8):7989–7996

Yeo AR (1999) Predicting the interaction between the effects of salinity and climate change on crop plants. Sci Hortic 78:159–174

Yeo AR, Flowers TJ (1986) The physiology of salinity resistance in rice (*Oryza sativa* L.) and a pyramiding approach to breeding varieties for saline soils. Aust J Plant Physiol 13:75–91

Young ND, Debellé F, Oldroyd GE, Geurts R, Cannon SB, Udvardi MK, Benedito VA, Mayer KF, Gouzy J, Schoof H, Van de Peer Y (2011) The *Medicago* genome provides insight into the evolution of rhizobial symbioses. Nature 480(7378):520

Zhen RG, Baykov AA, Bakuleva NP, Rea PA (1994) Aminomethylenediphosphonate: a potent type-specific inhibitor of both plant and phototrophic bacterial H+-pyrophosphatases. Plant Physiol 104:153–159

Zhen RG, Kim EJ, Rea PA (1997) Acidic residues necessary for pyrophosphate-energized pumping and inhibition of the vacuolar H⁺-pyrophosphatase by N, N′-dicyclohexylcarbodiimide. J Biol Chem 272:22340–22348

Index

A

ABC transporter gene, 175
Abiotic and biotic stress, 48, 207
Abiotic pressure, 79
Abiotic stress tolerance, 49, 86
Affymetrix Gene Chip Arrays, 39
Affymetrix Rice Genome Array, 56
Agricultural bioinformatics
    abiotic and biotic stresses, 32
    agronomic performance, 40
    applications, 30, 31
    breeding methods, 36
    climate change and population, 32
    crops, 29
    databases, 39, 41
    economic and environmental cost, 32
    genetic and molecular, 32
    genome sequencing, 32
    genomic data, 40
    genomics, 32
    GWAS, 40
    information and databases, 30
    knowledge and skills, 31
    modern technologies, 30
    multiple alignments, 34
    plant breeding, 35, 38
    plant genomics, 30
    QTL, 40
    substantial transformation, 36
    tool, 30
Agricultural genomics, 33
Agricultural proteomics research, 3, 10
Agricultural science
    ANOVA, 94
    authors' authorization, 94
    box plot, 96, 98

in CANONO and RStudio 3.5.1 software, 94
    chord diagram, 99, 100
    correlogram, 96, 97
    experimental design, 94
    graphical methods, 101
    hierarchical clustering, 101
    multivariate analyses and approaches, 94
    nuisance variables, 94
    PCA, 95
α-globin gene, 172
Amino methylene bisphosphonate (AMBP), 207
Amplified fragment length polymorphism
        (AFLP), 170, 181, 189
Analysis of variance (ANOVA), 94
Animal breeding proteomic research, 15
Applications
    bioinformatics, 143, 147
    computational methods, 150
    computational techniques, 149
    high-throughput bioinformatics
        approaches, 142
    remedial, 149
    therapeutic, 145
Aquaporins, 194
Arabidopsis Stress-Responsive Gene Database
        (ASRGD), 54
ArMet, 63
Artificial neural network (ANN), 124
Automated genotyping machines, 50
Automatic DNA sequencing, 50

B

Bacterial artificial chromosome (BAC), 160, 161
BAC-to-BAC approach, 160, 161
Bayesian method, 188

β-globin gene, 172
β-tubulin gene, 168, 171, 172
Bioinformatics, 1, 66, 115, 134
    applications, 143
    cellular processes, 151
    databases, 30
    drug designing, 141
    medicinal plants, 142, 150
    resources, 143–150
    softwares, 86
    system biology approaches, 110
    tools, 34
Bio-MassBank, 118
Bootstrapping, 183, 187
Box plot, 96, 98

C
Capillary electrophoresis-mass spectrometry
        (CE-MS), 129
Cardiovascular Disease Herbal Database
        (CVDHD), 149
cDNA libraries, 51
cDNA microarray, 162
ChemDP, 118
Cheminformatics, 33
China-US Million Book Digital Library
        Project, 144
Chord diagram, 99, 100
Combinatorial approach, 50
Comparative genetics, 33
Computational biology tools, 2
Computational metabolomics, 110, 134
Computational methods, 12
Conserved motifs, 197
Conserved segment (CS1), 198
Correlation Optimized Warping (COW)
        works, 121
1H-1H correlation spectroscopy (COSY), 129
Correlogram, 96, 97
Crop breeding, 40
Crop databases, 35
Crop improvement, 31, 32
Cytoglobin, 172

D
Database annotation quality
    DNA, 104
        plant cells and human brain, 104
Database management system (DBMS), 116
Database redundancy, 104, 105
Database search software programs/tools, 10
Data format conversion, 120

De novo sequencing, 10
Diammonium phosphate (DAP), 94
Difference gel electrophoresis (DIGE), 59
2,5-Dihydroxybenzoic acid (DHB), 126
3,5-Dimethoxy-4-hydroxycinnamic acid, 126
Disease resistance, 68
Diversity Arrays Technology (DArT), 39
DNA-chip technological innovation, 84
DNA microarrays, 81
Domestic livestock and animal proteomics, 14
Drought Stress Gene Database (DroughtDB), 55
Drug development, 142, 150, 151

E
Electron ionization-mass spectrometry
        (EI-MS) technique, 125
Enterotoxin and chlorera toxins, 107
Epigenomics, 110
ExPASy Proteomics site, 9
Expressed sequence tags (ESTs), 50, 51, 82, 145

F
Fast-atom bombardment (FAB), 126
Food production agrosystems, 79–80
Fourier transform infrared (FT-IR), 130
Free indication decay (FID), 129
Fruit breeding, 41
Fruit plant bioinformatics, 42
Full-length cDNA libraries, 34
Functional annotation, 51

G
Gas chromatography-mass spectrometry
        (GC-MS), 128
Gel-based proteomics, 7
Gene bank, 106
Gene expression analysis, 41
Gene expression network (GEN) analysis, 56
Gene finding, 32
Gene pyramiding, 34
Genealogical concordance phylogenetic
        species recognition (GCPSR), 170, 172
Genetic engineering and transgenic
        techniques, 80
Genetic markers, 34, 39
Genome comparative analyses, 33
Genome sequencing, 14
    BAC-to-BAC approach, 160
    cDNA microarray, 162
    drug development, and clinical trials,
        159–160

human genome project, 160
image analysis, 163
large-scale approach, 162
molecular biological research, 164
PCR, 162, 163
proteomics, 163
shotgun approach, 161, 162
technologies, 50
Genome sizes, 106
Genomes
chromosomal sizes, 105
chromosomes, 105
human, 105
size, 105
Genome-wide association studies (GWAS),
40, 131, 132
tools and molecular marker database,
37–38
Genome-wide molecular tools, 36
Genomic and protein sequence data, 103
Genomic approaches, 69
Genomic technologies, 85
Genomics, 30, 49, 110, 111, 113, 130, 141,
142, 145, 146, 151
Gentisic acid, 126

H
Heat map generating tools, 13
Heavy metal (HM) stress tolerance, 85
Hemoglobin, 172
Herb Ingredient's Target (HIT), 145
Heteronuclear multiple bond correlation
(HMBC), 129
1H-13C heteronuclear single-quantum
correlation (HSQC), 129
Hierarchical Cluster Analysis
(HCA), 123
Hierarchical clustering, 101
High-throughput omic tools, 85
High-throughput sequencing methods, 50
Host-pathogen interactions, 34
Human genome project, 49
Hybridization-based method, 56

I
Illumina GoldenGate Assay, 39
Image analysis software, 86
Indian Medicinal Plants Phytochemicals and
Therapeutics (IMPPAT), 145
Ingenuity Pathway Analysis (IPA), 13
Insect genomics, 35
Insertional mutagenesis, 84

J
Jackknifing, 187

K
Kazusa Omics Data Market, 118
Kishino-Hasegawa (KH) test, 188
KomicMarket, 118
Kyoto Encyclopedia of Genes and Genomes
(KEGG), 117, 144

L
Laboratory information management system
(LIMS), 86
Liquid chromatography-mass spectrometry
(LC-MS), 128
Liquid secondary ion mass
spectrometry (LSIMS), 126

M
M13 library, 161
Machine learning, 35
Madison-Qingdao Metabolomics Consortium
Database (MMCD), 117
Marker-assisted studies and selection
(MAS), 39
Mascot®, 88
Mass spectrometric data, 89
Mass spectrometry (MS), 3
Mass spectrometry search-related
software, 11
Mass spectroscopy (MS), 125
MassBase, 118
Massive parallel signature sequencing
(MPSS), 84
Mathematical science, 94
Matrix-assisted laser desorption ionization
(MALDI), 88, 126
Medicinal Plant Genomics Resource
(MPGR), 146
Medicinal plants
bioinformatics resources, 143–147,
149, 150
innumerable chemical compounds, 142
interdisciplinary research, 142
pharmaceutical companies, 142
synthetic drug development, 142
Metabolic flux analysis, 63
Metabolic QTL (mQTL), 131, 132
Metabolite profiling, 62
Metabolites, 110–115, 117–119, 122–125,
128–132, 134

Metabolome-genome-wide association studies
    (mGWAS), 131
Metabolomic profiling, 63
Metabolomics, 141, 142, 145, 147, 151
    agricultural, 132, 134
    analytical techniques, 125–130
    computational, 110
    data processing and pre-treatment, 119–122
    database, 116–118
    decipher complex, 110
    environmental, 115
    extraction method, 113
    fingerprinting, 115
    International Plant Metabolomics
        Congress, 110
    nontargeted, 113, 114
    OMICS, 130–132
    plants, 110
    quantification and identification, 113
    supervised methods, 123–125
    targeted, 114, 115
    unsupervised methods, 122, 123
Metabolomics Database of Linkoping
    (MDL), 117
Metabolomics databases, 63
Metabolomics techniques and methods, 62
Metabolomics technologies, 62
Metabolomics.jp, 118
Metagenomics, 141
Meta-genomics and evolutionary genomics, 32
Metagenomics and transcriptomics
    approaches, 34
Metal geometry, 205
METLIN, 117
Metropolis-coupled Markov chain Monte
    Carlo (MCMCMC), 168
Microarray technology, 33
MicroRNA (miR), 63
Mini-arrays, 51
Mitochondrial DNA (mt DNA), 175
Mitochondrial ribosomal RNA (mt rRNA),
    170, 171
Molecular biology, 81
Molecular DNA marker, 34
Molecular phylogeny, 175, 183, 189
Multivariate approach, 94
Mutant analysis, 34
Myoglobin, 172

N
National Center for Biotechnology
    Information (NCBI), 56, 82
National Science Foundation (NSF), 50

Natural Product Activity and Species Source
    database (NPASS), 149
Naturally occurring Plant-based Anti-cancer
    Compound-activity-Target database
    (NPACT), 149
Neuroglobin, 172
Next-generation DNA sequencing application, 56
Next-generation sequencing (NGS), 50, 145
Non-model-pathogen interaction, 4–6
Nontargeted metabolomics, 113, 114
Nuclear Magnetic Resonance (NMR)
    Spectroscopy, 129–130
Number of leaves per plant (NoLPP), 94, 95

O
Omics approaches, 48
Omics platforms, 51
OMICS techniques, 130
Organelle expression proteomics, 9

P
Paralogous genes, 172
Partial least squares (PLS) regression, 123
PASmiR, 66
Pathogen-host interaction, 69
Pathogen proteins, 7
Pathogen Receptor Genes Database
    (PRGDB), 54
PCR method, 162, 163
Peptide/protein sequence, 8
Pharmacogenomics, 141
Phenomics, 110
Phylogenetic tree
    ABC transporter gene families, 175
    Bayesian method, 188
    bootstrapping, 183, 187
    constructing, 182, 183
    data, 181
    description, 168
    jackknifing, 187
    KH test, 188
    MCMCMC, 168
    rRNA sequences, 175
    SH test, 188, 189
Phylogeny, 195
Phylogeny packages, 183–186
Physiological and molecular cell processes, 10
Plant breeding, 35
Plant Expression Database (PLEXdb), 146
Plant genomic period, 50
Plant genomics, 30
Plant genomic sequences, 32

Plant genomics study, 50
Plant Metabolic Network (PMN), 117, 147
Plant metabolites
 by-products, 111
 compound, 111
 enzyme-catalyzed reactions, 111
 food and medicine, 111
 primary metabolites, 112
 secondary metabolites, 112, 113
Plant Metabolome Database (PMDB), 117
Plant MicroRNA Database (PMRD), 66
Plant Stress Gene Database (PSGD), 51
Plant Stress Protein Database (PSPDB), 61
Plant Stress Proteome Database
 (PlantPReS), 61
Plant Stress RNA-Seq Nexus (PSRN), 56
Plantmetabolomics.org (PM), 63, 116
Plant-pathogen interactions, 69
Plants, 111–113, 116, 117, 128, 131–134
PmiRExAt, 66
Polymerase chain reaction (PCR), 83
Principal component analysis (PCA), 123
Protein interactions, 11
Protein-protein interactions (PPIs), 12, 68
Proteogenomics, 61
Proteoinformatics, 2, 7, 8
 bioinformatic tools, 7
 computational biology, 1
 crops and animals, 8
 DNA/RNA sequences, 2
 functional analysis, 1
 genoinformation, 8
 genome-based information, 8
 host-pathogen interactions, 7
 host-pathogen systems, 2
 immunity mechanism, 3
 multidisciplinary research, 3
 NCBInr database, 10
 non-model organism, 3
 online databases, 9
 phenotype and genotype, 2
 phylogeny tools, 2
 plant/animal quality, 2
 protein separation, 10
 proteoinformation, 7
 stress, 9
Proteome, 58
 analysis, 106
 characterization, 59
 datasets, 59
 mass spectrometry, 59
 and phosphoproteomics, 59
 profiling, 81
 technology, 58

Proteomics, 68, 80, 83, 110, 129, 130, 141,
 145, 147, 151, 163
 abiotic stress in plants, 80
 analysis, 13
 approach, 85
 cDNA-AFLP, 83
 climate change, 79
 DDRT, 83
 functional genomics, 81
 genomics, 80
 insertional mutagenesis, 84
 mRNA, 82
 protein evaluation, 81
 SAGE, 84
Proton pump, 194, 206, 207

Q
QlicRice, 54
Quantitative Composition-Activity
 Relationship (QCAR), 149
Quantitative PCR analyses, 56
Quantitative Structure–Activity Relationship
 (QSAR), 149
Quantitative trait loci (QTL), 36, 40, 131

R
Random forests (RF), 124
Research Collaboratory for Structural
 Bioinformatics (RCSB), 144
ReSpect, 118
RiceArrayNet (RAN), 58
Rice SRTFDB, 54
RNA/gene expression profiling, 55
RStudio 3.5.1 software, 94

S
Search Tool for the Retrieval of Interacting
 Genes/Proteins (STRING), 12
Sequence-based profiling, 65
Sequest®, 88
Serial analysis of gene expression (SAGE), 84
Shimodaira-Hasegawa (SH) test, 188, 189
Shotgun approach, 161, 162
Silico genomics technology, 34
Sinapinic acid, 126
Single-stranded confirmation polymorphism
 (SSCP), 170, 171, 181, 189
Standard Metabolic Reporting Structures
 (SMRS), 63
Statistical analysis, 107
Stress-related databases, 59

Stress-responsive Transcription Factor (STIF)
    algorithm, 53
Stress-Responsive Transcription Factor
    Database (STIFDB V2.0), 53
Stress-Responsive Transcription Factor
    Database (STIFDB2), 53

T
Targeted metabolomics, 114, 115
TCM Information Database (TCM-ID), 145
Tetramethylsilane (TMS), 129
The Arabidopsis Information Resource
    (TAIR), 54, 146
The Institute for Genomic Research
    (TIGR), 161
Therapeutic Target Database (TTB), 145
Tiling arrays, 56
1H-1H total correlation spectroscopy
    (TOCSY), 129
Traditional Chinese Medicine (TCM)
    system, 144
Transcriptome analysis, 56
Transcriptome Encyclopedia of Rice
    (TENOR), 58
Transcriptomics, 48, 110, 130, 141, 145,
    147, 151
  environmental conditions, 55
  GEN, 56
  hybridization-based method, 56
  next-generation DNA sequencing
      application, 56
  PSRN, 56
  quantitative PCR analyses, 56
  RNA expression profile, 55
  RNA/gene expression profiling, 55
  RNA-Seq database, 56
  TENOR, 58
  tiling arrays, 56

Tree construction methods, 182
2-D gel image analysis software, 87

U
Unweighted pair group method with
    arithmetic mean (UPGMA),
    170, 182
The US National Library of Medicine
    (NLM), 144

V
Vacuolar proton-translocating inorganic
    pyrophosphatases (VPPases)
  global climate, 193
  isoforms, 194
  land plants, 197
  molecular phylogeny, 195–196
  phosphate binding sites, 207
  potassium ion, 206
  ribbon structure, 204
  salinity stress, 194
  sequence positions, 201
  solubilization and purification, 205
  structural model, 196
  3D structure, 201
  tobacco, 195
  topological analyses, 199
  transmembrane helices, 200–202
  vacuolar proteins, 194
Vibrational spectroscopy, 130
Virtual screening, 149
Vitamin A, 34

W
Whole genome array (WGA), 146
Whole-genome sequencing, 32, 33

CPI Antony Rowe
Eastbourne, UK
November 15, 2019